CONDENSED CHEMISTRY

GUY WALLER

Radley College, Radley, Oxfordshire

HODDER AND STOUGHTON
LONDON   SYDNEY   AUCKLAND   TORONTO

British Library Cataloguing in Publication Data

Waller, Guy
  Condensed chemistry.
  1. Chemistry
  1. Title
  540   QD33

ISBN 0 340 35217 5

First printed 1985

Copyright © 1985 Guy Waller

All rights reserved. No part of this publication may be reproduced or transmitted in any form or by any means, electronic or mechanical, including photocopy, recording, or any information storage and retrieval system, without permission in writing from the publisher.
Typeset in 10/12 pt Univers Medium
Printed and bound in Great Britain for
Hodder and Stoughton Educational,
a division of Hodder and Stoughton Ltd.,
Mill Road, Dunton Green, Sevenoaks, Kent TN13 2YD,
by J. W. Arrowsmith Ltd., Bristol BS3 2NT

Contents

1 Particles 1
1.1 atoms 1.2 bonding
1.3 molecules 1.4 ions

2 Phases 29
2.1 gas 2.2 liquid 2.3 solid

3 Physical equilibrium 49
3.1 phase changes in pure substances
3.2 phase changes in solution

4 Chemical change 58
4.1 enthalpy changes
4.2 Born–Haber cycles

5 Reaction rates 69
5.1 measuring rates
5.2 collision theory
5.3 multi-step reactions

6 Chemical equilibrium 80
6.1 the equilibrium law
6.2 proton transfer
6.3 electron transfer

7 Electrochemistry 100
7.1 conductivity 7.2 electrolysis

8 The chemistry of the metals 109
8.1 Group I 8.2 Group II
8.3 aluminium
8.4 the transition metals
8.5 chromium and iron
8.6 the B-metals

9 The chemistry of the non-metals 130
9.1 Group VII 9.2 Group VI
9.3 Group V 9.4 Group IV
9.5 hydrogen

10 Industrial chemistry 165
10.1 metal extraction 10.2 alkalis
10.3 acids 10.4 petrochemicals

11 Reactivity of organic compounds 178
11.1 mechanism
11.2 nucleophilic reactions
11.3 electrophilic reactions

12 Classes of organic compounds 188
12.1 alkanes 12.2 alkenes
12.3 aromaticity 12.4 halides
12.5 alcohols and phenols
12.6 amines
12.7 aldehydes and ketones
12.8 carboxylic acids
12.9 acid chlorides and anhydrides
12.10 esters and amides

13 Organic reagents: a summary 231
13.1 diagnostic reagents
13.2 synthetic reagents

14 Stereochemistry 239
14.1 isomerism
14.2 stereospecificity and stereoselectivity

Index 250

Preface

Chemistry is a broad subject. Any author who attempts to condense a study of chemistry into two hundred and fifty six pages faces a daunting prospect! The text must be clear and readable, but present the subject in a visually interesting way. It must be friendly to those who seek information from it, but have sufficient structure to show how the jigsaw fits together.

In writing *Condensed Chemistry*, I have tried to keep all this in mind. Wherever possible, information is presented in the form of a diagram, or as a series of numbered points, and shaded boxes are used to pick out important concepts, laws and definitions. Descriptive inorganic chemistry is treated in terms of the structure, acid-base and redox properties of the element concerned, while descriptive organic chemistry is given a mechanistic flavour (which I hope is not too strong). With fifty-one short sections in fourteen chapters, the text is quite highly structured, and this should assist the reader when seeking a specific piece of information. There is also a full index.

Condensed Chemistry is not designed with any particular course in mind. Most teachers like to choose their own path through an A-level course. Whatever the route, I hope that *Condensed Chemistry* may offer their students the supplementary back-up needed, particularly as they draw near to the end of the course and look back to see how it all fits together. Many of my students have helped to knock some of the wrinkles out of the text, and I am greatly indebted to them. I would also like to thank John Clarke for his invaluable advice and work on the manuscript, and also the staff of Hodder and Stoughton for their encouragement, patience and thoroughness. Finally, my biggest thanks are to my wife, for whom this short book was written, for her tolerance.

Guy Waller

1 Particles

1.1 ATOMS

Composition and stability

In 1897, radioactivity was accidentally detected by Becquerel. He found the image of a key on a previously unexposed photographic plate which he had stored next to a sample of a radium salt in a closed drawer. Shortly after this discovery, a sample of radium was shown to be able to emit a stream of fast-moving, positively-charged particles which were called α-particles. Rutherford, Geiger and Marsden designed a series of experiments in which a very finely beaten piece of gold foil was used as a target in front of a radium source. The α-particles were scattered by the target and detected by a zinc sulphide screen which glowed when struck by an α-particle.

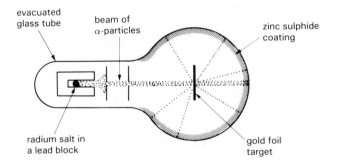

These 'scattering experiments' (1910–1911) showed that:

1 α-particles passed through the gold target with almost no hindrance at all;
2 a few deflections occurred and these were detected at all angles around the target.

2 Condensed Chemistry

They concluded that:

1 an atom was largely empty space;
2 the mass of an atom was concentrated in a tiny, positively-charged centre (the nucleus);
3 there were very light, negatively-charged particles in orbit around the nucleus (electrons).

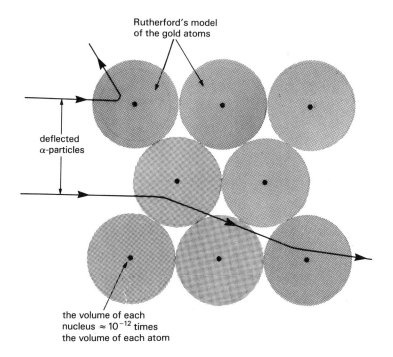

The existence of neutrons was first implied by the results of mass spectrometry (page 21). Aston (1918) showed that a sample of a pure element contained atoms of different mass. It was suggested that there was a third type of neutral, subatomic particle whose mass was approximately equal to that of a proton. The number of these particles in the nucleus of an atom vary.

> The atomic number, Z is the number of protons in the nucleus of an atom; the mass number, A is the number of protons and neutrons in the nucleus; two isotopes are atoms with the same atomic number but different mass numbers.

An atom's composition is written in the form $^A_Z X$ where X is the symbol for the element, A is the mass number and Z is the atomic number. For example, there are two common isotopes of chlorine, $^{37}_{17}Cl$ and $^{35}_{17}Cl$, and one has two more neutrons per atom than the other.

Nuclear stability. The nucleus of an atom is held together by extremely strong, but short-range forces between the particles present (nucleons). For a particular nucleus, it is known as the nuclear binding force. It is opposed by the electrostatic forces of repulsion between the protons. Sometimes the electrostatic repulsive forces exceed the nuclear binding force and the nucleus undergoes fission. In bulk terms, this process is called radioactive decay. There are three main types of radioactivity.

1. α-decay: an unstable nucleus ejects an α-particle which is the nucleus of a helium atom.

    $^{228}_{90}\text{Th} \longrightarrow {}^{224}_{88}\text{Ra} + {}^{4}_{2}\text{He}$

2. β-decay: an unstable nucleus ejects a β-particle which is usually an electron. This takes place most often as a result of a neutron disintegrating to form a proton and an electron.

    $^{32}_{15}\text{P} \longrightarrow {}^{32}_{16}\text{S} + {}^{0}_{-1}\text{e}$

3. γ-decay: a nucleus can only possess certain fixed amounts of energy. It has different energy levels because nuclear energy is quantized like electronic energy (see page 5). When a nucleus in a high energy level (E_2) returns to a lower level (E_1), it emits very high frequency radiation called γ-rays. $(E_2 - E_1) = \Delta E = h\nu$, where ν is the frequency of the γ-rays emitted and h is Planck's constant.

    $^{80}_{35}\text{Br}^* \longrightarrow {}^{80}_{35}\text{Br} + h\nu_\gamma$

radiation	penetration	effect of magnetic field
α-rays	stopped by 1 mm of aluminium	deflected in the opposite direction to β-rays
β-rays	stopped by 5 mm of aluminium	deflected about twice as much as α-rays
γ-rays	stopped by 15 cm of lead	no deflection

All three types of radiation are detected by a Geiger–Muller tube.

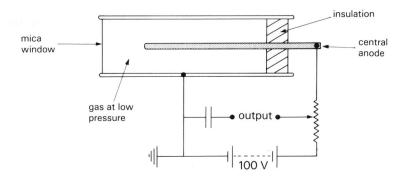

The radiation ionizes the gas present at low pressure inside the tube. The ions are attracted to the electrodes and generate a small pulse of current which is amplified and fed to a recorder. The reading in counts per second (becquerels) given by the recorder is a measure of the amount of radioactive substance present in the sample.

Rates of radioactive decay follow first order kinetics (see page 70). The concentration of radioactive material is determined at various times by recording the becquerels detected after each interval.

Atomic spectra

When a gaseous sample of an element is introduced at low pressure between two electrodes of high potential difference, light energy is emitted. On analysing the light with a diffraction grating, it is found that the frequencies do not form a continuous range (as is the case for white light or sunlight). For example, for a hydrogen discharge tube:

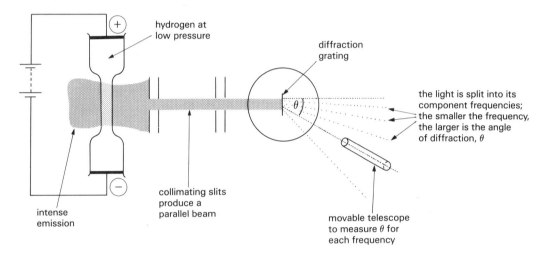

If a photographic plate is used instead of a movable telescope, the film is exposed as shown below (only one small range of the spectrum is shown).

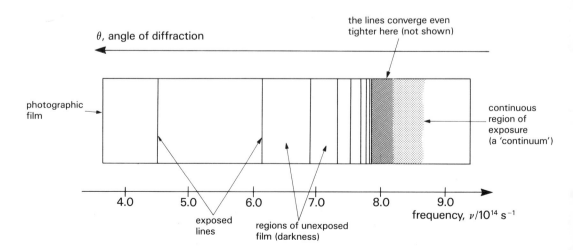

The converging pattern of the lines in the atomic spectrum suggests two important ideas:

1 that orbiting electrons can only possess certain fixed amounts (quanta) of energy: orbitals are at one 'energy-level' or another;
2 that energy-levels gradually become closer in value as the radius of the electronic orbital increases.

These two conclusions explain the observations about atomic spectra as follows.

1 An unexcited atom has electrons in the lowest available energy-level (the atom is in its 'ground state').
2 In the electric field, an electron takes in energy and moves to a higher energy-level. For example, an electron is excited from E_1 to E_5 in the diagram below.

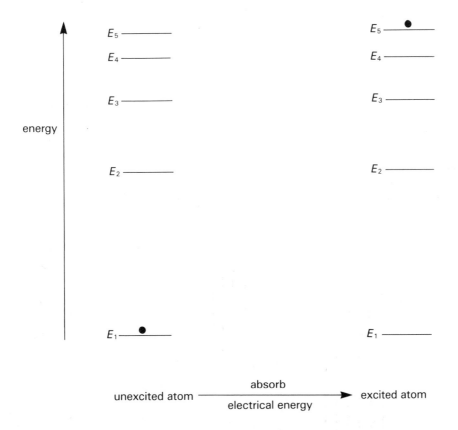

3 In returning to a lower energy-level (for example, E_5–E_2) the electron gives out the excess energy as light.
4 The frequency (ν) of the emitted light can only have a certain specific value which is determined by the difference between the two energy-levels involved, ΔE. In the example overleaf, $\Delta E = E_5 - E_2$. $\Delta E = h\nu$ where h = Planck's constant.

6 Condensed Chemistry

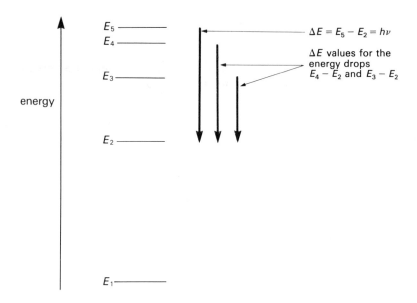

5 The frequencies of the emissions get closer in value because the energy-levels within the atom are getting closer in value.
6 There are regions of continuous light produced by the merging of spectral lines whose frequencies are so close together in value that they cannot be distinguished. A region of this sort is called a continuum.
7 Dependent on which level the electrons return to, there are a number of possible series of lines:

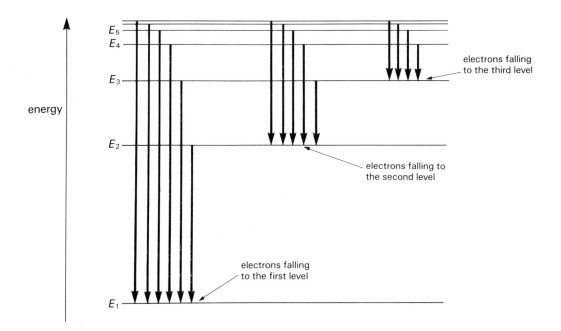

8 There are only three visible lines on the hydrogen spectrum because all the other emissions are in the infra-red or ultra-violet range. The visible lines are due to electrons returning to the second energy-level.

The start of the continuum of highest frequency in the hydrogen spectrum represents the energy drop when an electron in the very last energy-level falls back to the first level. This energy difference is also equivalent to the energy needed to ionize a hydrogen atom. For a mole of atoms, it is called the ionization energy (*IE*).

> The molar ionization energy of an element is the energy required to form a mole of gaseous cations from a mole of gaseous atoms, each atom losing one electron at 298 K:
>
> $X_{(g)} \longrightarrow X^+_{(g)} + e^- \qquad \Delta E = IE$

Ionization energies are either determined from spectral values or by 'electron impact' measurement.

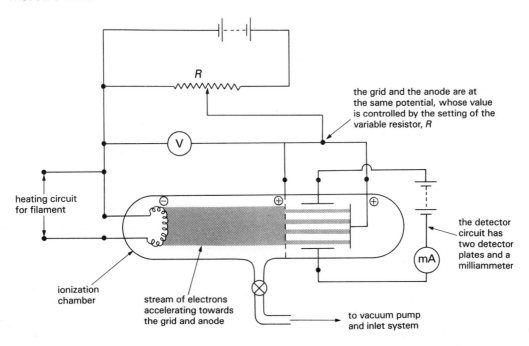

A gaseous sample of an element is introduced at low pressure into the evacuated ionization chamber. The cathode is heated and a stream of electrons is accelerated towards the grid and anode. Because the grid and anode are at the same potential, the electrons that pass through the grid then travel at a constant velocity (and hence with constant energy) towards the anode. Between the grid and anode, an electron may collide with a sample atom, and if it has enough energy, it may knock an outer-shell electron out of the atom on impact.

The energy required to knock out an electron is equivalent to the ionization energy. If any ions are produced they are attracted to electrodes of the detector circuit and a current flows. A plot of detector current against applied voltage has the following form.

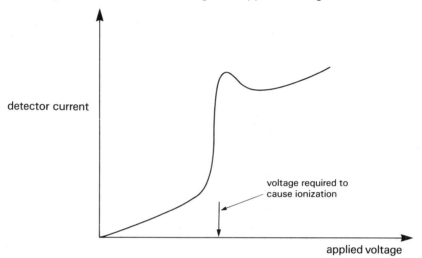

The applied voltage required to cause a sharp increase of the detector current is a measure of the energy that the electrons need to bring about ionization on impact. For example, when the ionization chamber contains sodium vapour, an applied voltage of 5.12 V causes a sharp increase. The energy of a mole of electrons accelerated through this potential difference equals $(5.12 \times q \times L)$ J, where q = the charge of an electron in coulombs and L is the Avogadro constant. Hence the ionization energy of sodium is given by: $(5.12 \times 1.602 \times 10^{-19} \times 6.022 \times 10^{23}) = 494$ kJ mol^{-1}.

Electronic structure

Two important generalizations can be made about the nature of electrons.

1. Electrons have both wave-like character (an electron can be diffracted) and particle-like character (an electron beam can be made to drive a light paddle-wheel).
2. An electron in an atom is confined to one of a number of orbitals which are grouped in shells going out from the nucleus.

> An orbital is defined as a volume of space in which it is highly probable that a particular electron will be found.
> A shell is a group of orbitals of approximately equal radial distribution from the nucleus.

These generalizations are further extended by the mathematical treatment of electronic wave-motion:

1. only two electrons can be fitted into each orbital (a form of the Pauli exclusion principle, see page 11);
2. only n^2 orbitals can be fitted into the nth shell;

3 the nth shell has n subshells, each subshell containing a set of orbitals of equal energy and similar shape. Subshells are lettered either s, p, d or f as shown below.

subshell	orbital shape	number of orbitals in each subshell
s	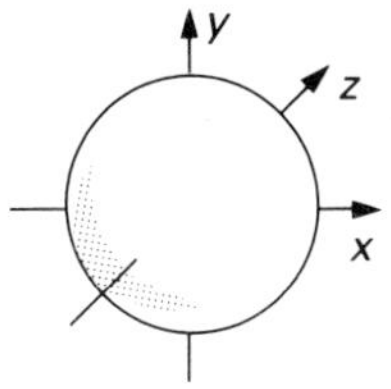	1
p	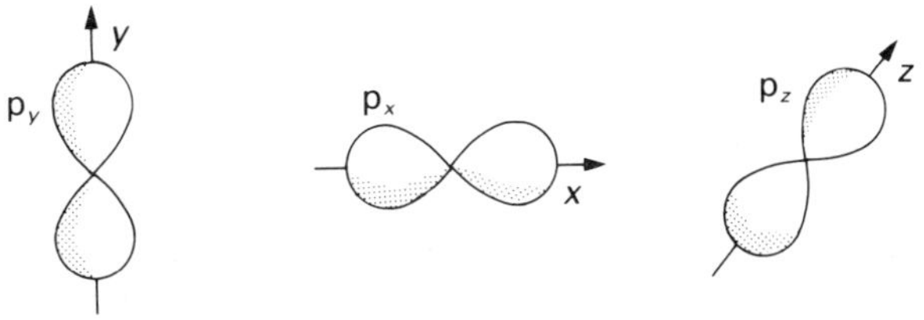 p_y, p_x, p_z	3
d	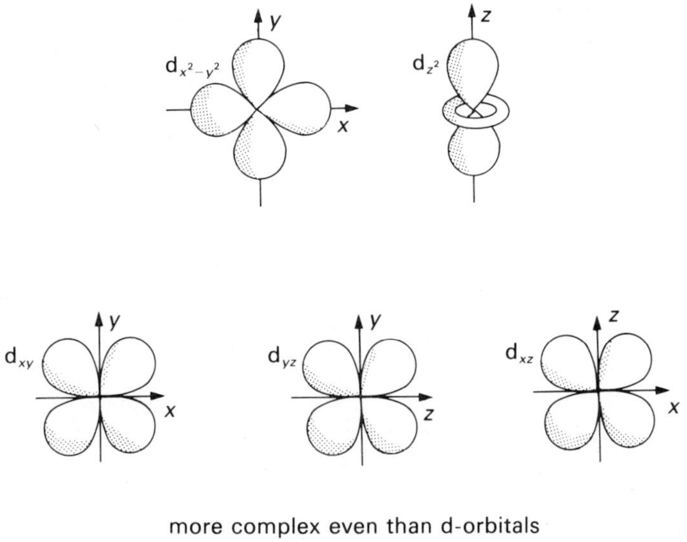 $d_{x^2-y^2}$, d_{z^2}, d_{xy}, d_{yz}, d_{xz}	5
f	more complex even than d-orbitals	7

For example, the 3rd shell has **3** subshells and 3^2 orbitals = **9** orbitals. It has an s-subshell, a p-subshell and a d-subshell; this is a total of $(1 + 3 + 5) = 9$ orbitals.

4 In building up the electronic structure of an atom, electrons are placed successively into the orbital of lowest available energy (the *aufbau principle*).

The order of filling orbitals can be remembered by using the triangle shown below. The electronic configuration of phosphorus is given beside it to illustrate all these ideas.

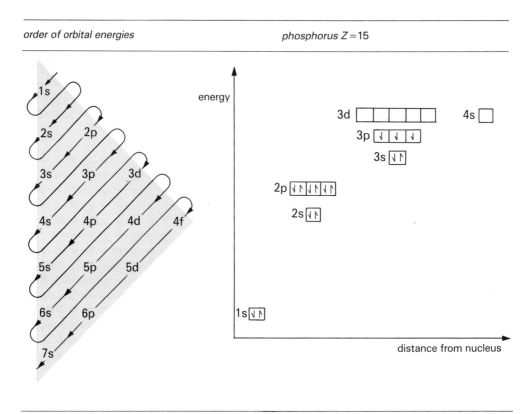

order of orbital energies   *phosphorus Z = 15*

In the electronic configuration shown above ($1s^2\, 2s^2\, 2p^6\, 3s^2\, 3p^3$), it should be noted that the last three electrons go one each into the three orbitals of the outer p-subshell. A half-filled orbital is of lower energy than a paired orbital because electron repulsion is lower when the orbital is half-full (the *Hund principle*).

Quantum numbers. The solutions of the wave equation of electronic motion suggest that electronic energy is confined to particular values ('quantized'). In the absence of a magnetic field, the energy of an electron is controlled by two factors: its distance from the nucleus and its angular momentum. The permitted wave functions are, therefore, governed by two 'quantum numbers', n and l. Both n and l may only have integral values, and these values indicate the energy of the electron concerned: as their values increase, so does the energy-level of the electron.

The first number, n, is called the principal quantum number and is associated with the radial distribution of an electronic orbital from the nucleus; it corresponds well with the shell number. The second number, l, is called the angular momentum quantum number and is associated with the shape of the orbital concerned. The value of l denotes the type of orbital: $l = 0$, s-orbital; $l = 1$, p-orbital; $l = 2$, d-orbital etc.

permitted values: $n = 1, 2, 3 \ldots \infty$
$l = 0, 1, 2 \ldots (n-1)$

value of n:	1	2	2	3	3	3	4	4	4	4	5	5	5	etc.
value of l:	0	0	1	0	1	2	0	1	2	3	0	1	2	

1st shell · 2nd shell · 3rd shell · 4th shell · 5th shell

low energy — ELECTRONIC ENERGY-LEVEL — high energy

In a strong, homogeneous magnetic field, it is observed that the number of permitted energy-levels increases sharply. Two separate effects are responsible for this.

1. An orbital is characterized not only by its radial distribution and shape, but also by its orientation in space. In the absence of a magnetic field, there is no distinction in energy between any two orientations. However, in a strong external field, an orientation in the direction of the field is of different energy from one that points across it, and so on. A magnetic quantum number m_l denotes the number of possible orientations for any given value of l. There are $(2l + 1)$ possible values of m_l for example,

when $l = 1$ $m_l = -1, 0$ or 1
$l = 2$ $m_l = -2, -1, 0, 1$ or 2 etc.

2. An electron also appears to have a property that is best understood as 'spin'. Since a spinning charge sets up its own magnetic field, in an external field there is a difference in energy between the clockwise and the anticlockwise spin-state. A spin quantum number m_s denotes the direction of electronic spin. m_s can have a value either $+\frac{1}{2}$ or $-\frac{1}{2}$.

To summarize: four quantum numbers, n, l, m_l and m_s are needed for a full description of the energy of an electron in a particular orbital. The Pauli exclusion principle states that no two electrons can have the same set of quantum numbers, and this leads to the following interpretation of shell structure.

	n	l	m_l	m_s	
first shell	1	0	0	$-\frac{1}{2}$	$1s^2$
	1	0	0	$+\frac{1}{2}$	
second shell	2	0	0	$-\frac{1}{2}$	$2s^2$
	2	0	0	$+\frac{1}{2}$	
	2	1	-1	$-\frac{1}{2}$	$2p_x^2$
	2	1	-1	$+\frac{1}{2}$	
	2	1	0	$-\frac{1}{2}$	$2p_y^2$
	2	1	0	$+\frac{1}{2}$	
	2	1	1	$-\frac{1}{2}$	$2p_z^2$
	2	1	1	$+\frac{1}{2}$	

1.2 BONDING

Bond energy

When two atoms come into contact, a number of electrostatic forces are set up between them. The two electron-clouds experience forces of repulsion from each other and there is also repulsion between the nuclei: balancing this repulsion are forces of attraction between the nuclei and the electrons. It is, therefore, possible to identify three different results arising from the balance of these forces.

1 The forces of repulsion outweigh those of attraction.
2 The forces of attraction outweigh those of repulsion.
  (i) The electron-attracting power of each nucleus is approximately equal.
  (ii) The electron-attracting power of one nucleus is much greater.

The three cases are illustrated by the curves on the facing page. Curve a) illustrates case 1 and curves b) and c) illustrate 2(i) and 2(ii).

A measure of the attractive force that a nucleus has for its own electrons is given by the ionization energy (page 7). However, there is a similar energy term that describes the attractive force of a nucleus for electrons other than its own. It is called the electron affinity (EA).

> The molar electron affinity of an element is the energy change when a mole of gaseous atoms accept a mole of electrons, each atom becoming a single-charged gaseous anion at 298 K.
>
> $X_{(g)} + e^- \longrightarrow X^-_{(g)} \qquad \Delta E = EA$

In a pair of bonded atoms, it is necessary to assess the attractive power that each nucleus exerts over both its own and the other atom's electrons. Electronegativity is a term which takes account of both effects. It has a scale of values ranging from 0.7 to 4.0.

> The electronegativity of an element is a measure of the electron-attracting power that an atom of the element exerts on the bonding and non-bonding electrons present in its outer shell.

Covalent bonding is most likely to occur between two elements of similar and high electronegativity (both greater than 2.0). On curve b) opposite, the difference, D, between the potential energy at the minimum and the potential energy of the isolated atoms is called the bond dissociation energy. Ionic bonding is most likely to occur between two elements of widely different electronegativity. On curve c), the difference in potential energy, U, is called the lattice energy.

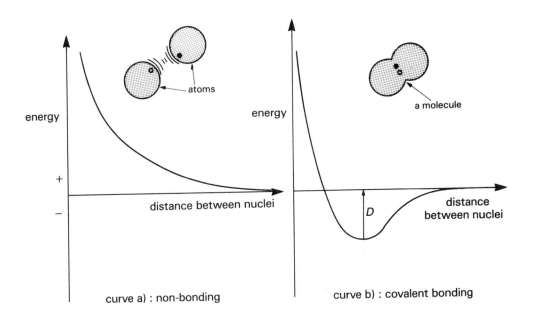

curve a) : non-bonding

curve b) : covalent bonding

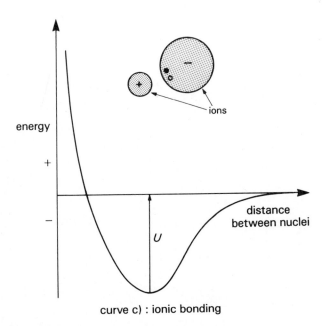

curve c) : ionic bonding

covalent model
The bond dissociation energy, D, is the energy required to break a mole of bonds when the process is carried out in the gas phase at 298 K and 0.1 MPa pressure.

ionic model
The lattice energy, U, is the energy required to break up a mole of ionic lattice into separate ions an infinite distance apart in the gas phase at 298 K and 0.1 MPa pressure.

These energy terms are illustrated by the bonding models on pages 64 and 65. Some examples are shown below.

$D/kJ\,mol^{-1}$		$U/kJ\,mol^{-1}$	
Cl—Cl	242	Na^+Cl^-	771
O=O	496	$Mg^{2+}O^{2-}$	3889

A fuller description of the factors favouring covalency is given overleaf.

Covalent bonding

When two atomic orbitals overlap, two 'molecular' orbitals form because the electrons in the orbitals come under the influence of both nuclei. The average energy of these molecular orbitals equals that of the atomic orbitals, one molecular orbital having a higher energy-level while the other has a lower energy-level. If there is only a single pair of electrons to be accommodated, the aufbau principle suggests that both electrons go into the lower molecular orbital. There is thus a lowering in the potential energy of the system, and this corresponds to the bond energy. For example, consider the case of two hydrogen atoms when their 1s-orbitals overlap:

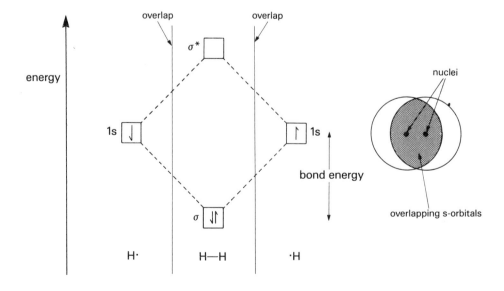

The upper molecular orbital is called an anti-bonding orbital (shown by *) because its energy-level exceeds that of the separate atoms. The lower molecular orbital is a bonding orbital.

Orbital overlap. There are two types of overlap.

1 σ-overlap: the overlap of a single lobe of one orbital with a single lobe of another, for example, 1s and 1s or,

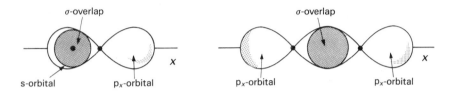

2 π-overlap: the simultaneous overlap of two different lobes of one orbital with two lobes of another, for example,

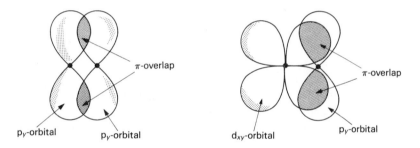

π-overlap can only occur between the orbitals of atoms already joined by σ-overlap. When the potential energy of the system decreases as a result of the presence of electrons in σ-orbitals or π-orbitals, the terms σ-bond or π-bond are used. For example, the double bond in an ethene molecule is made up from one σ-bond and one π-bond.

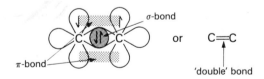

Electron-pair type. The above diagram illustrates that there are two distinctly different types of electron-pair.

1 A σ-pair occupies a volume along the line of centres of the bonded atoms.
2 A π-pair occupies a volume above and below the σ-pair.

There is a third type of electron-pair commonly found in a molecule or molecular ion. It is called a 'lone pair', and is an outer-shell pair of electrons under the influence of only one nucleus. For example, a carbonyl group (shown overleaf) contains a σ-pair, a π-pair and two lone pairs.

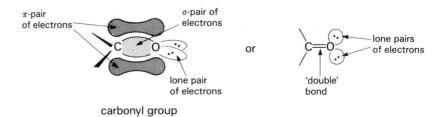

carbonyl group

Degree of covalency

The covalent and ionic models of bonding represent two extreme cases. The bonding present in a particular compound contains characteristics of both extremes, and the terms 'degree of covalency' or 'percentage ionic character' are used to describe the relative balance.

Fajans summarized the factors that lead two atoms to bond together with a low degree of covalency.

> Fajans' rules state that ionic bonding is most likely between two atoms when:
>
> 1 the cation likely to be formed is large;
> 2 the anion likely to be formed is small;
> 3 the charges on the ions likely to be formed are low.

When these three criteria are not met, the cation polarizes the anion and a high degree of covalency results. In other words, a small highly-charged cation has considerable 'polarizing power' while a large, highly-charged anion has considerable 'polarizability'.

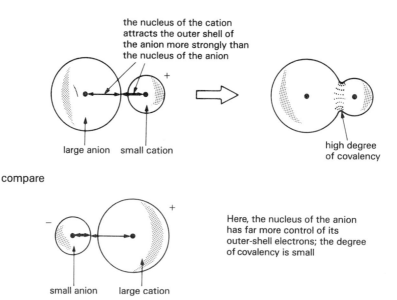

As an illustration of Fajans' rules, they are applied (below) to the bonding in potassium fluoride and tetrachloromethane.

KF	CCl_4
1 large cation K^+	1 minute cation C^{4+}
2 small anion F^-	2 medium-sized anion Cl^-
3 low charges	3 highly-charged cation $4+$
∴ low degree of covalency	∴ high degree of covalency
Sensible use of the ionic model	Better to use the covalent model

1.3 MOLECULES

Molecular shape

The σ-bonds and lone pairs in the outer shell of an atom are directional. They repel one another around the atom until a position of equilibrium is reached. A π-bond, however, does not significantly alter the shape of a molecule because it surrounds a σ-bond.

To predict the shape of a molecule (or a molecular ion), Sidgwick and Powell's rules are useful.

1. Each σ-pair and lone pair achieves a position as far from the others as possible.
2. Lone pairs and double-bonded pairs repel more than single-bonded pairs; wherever possible they take up equatorial rather than axial positions. For example, in the 5 co-ordinate arrangement:

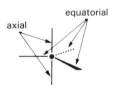

The rules are applied in one of two ways.

For particles without oxygen atoms

1. Count the number of outer-shell electrons in the central atom.
2. Add one for each bonding atom.
3. Add one for each negative charge (if the particle is anionic) or subtract one for each positive charge (if the cationic).
4. Divide the total by two to give the number of electron-pairs.
5. Subtract the number of bonded atoms to find the number of lone pairs present on the central atom. For example, the shape of ICl_4^+ is worked out as follows:
   (i) I has 7 electrons,
   (ii) there are 4 bonding atoms,
   (iii) one electron has been lost,
   (iv) $(7 + 4 - 1) \div 2 = 5$ pairs,
   (v) four are bonding pairs, therefore one is a lone pair.

the lone pair occupies an equatorial site

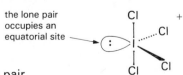

For particles with oxygen atoms. These are usually oxyanions containing double-bonded oxygen atoms and it is, therefore, necessary to determine the number of double-bonded atoms before the shape can be predicted. This is achieved by making the following assumptions:

1. for each negative charge, there is one, —Ö: group.
2. for each hydrogen atom, there is one, —OH group.
3. all the remaining oxygen atoms are double-bonded. For example, the shape of HSO_3^- is worked out below.

   (i) One negative charge

   ∴ one —Ö: group

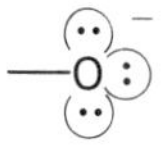

   (ii) One hydrogen atom

   ∴ one —OH group

   (iii) Only one oxygen atom is left

   ∴ one ═O: group

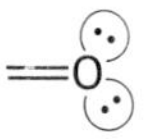

Sulphur has six outer-shell electrons and, because it is forming a total of four bonds, there is one lone pair.

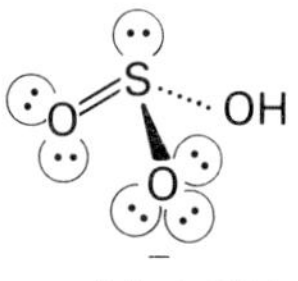

hydrogensulphate(IV) ion, HSO_3^-

Molecular lattices

Molecular lattices are held together by one or more of the following of inter-molecular forces:

1. van der Waals' forces   5–20 kJ mol^{-1};
2. dipole–dipole forces   10–30 kJ mol^{-1};
3. hydrogen bonds   20–50 kJ mol^{-1}.

Van der Waals' forces are caused by the instantaneous dipoles set up inside each molecule as a result of a temporarily uneven electron distribution. The effect is illustrated below.

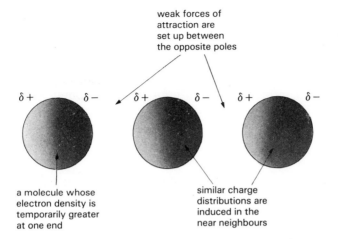

Dipole–dipole forces are present in all molecular lattices where the bonding electrons are shared unequally within the covalent bonds present. This separation of charge along each bond is the result of an electronegativity difference between bonded atoms. For example,

$$\overset{\delta+\;\;\;\delta-}{\text{H—Cl}} \qquad \begin{array}{l}\text{Electronegativities}\\ \text{H}=2.1 \quad \text{Cl}=3.0\end{array}$$

$$\overset{\delta-\;\;\delta+\;\;\delta-}{\text{O=C=O}} \qquad \text{C}=2.5 \quad \text{O}=3.5$$

Attraction takes place between the opposite poles of neighbouring molecules. For example, the structure of dry ice is as follows.

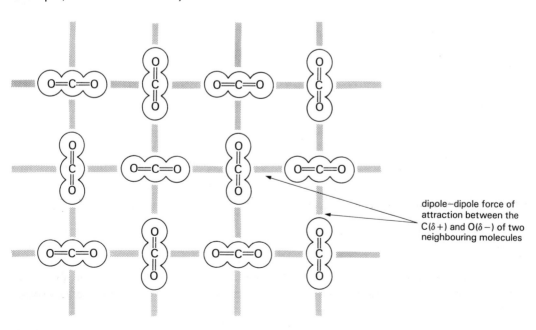

Hydrogen bonds are particularly powerful dipole–dipole forces involving covalently-bonded hydrogen atoms. When a hydrogen atom bonds to a small highly-electronegative atom, it loses partial control of its only electron. Because the hydrogen nucleus is a single proton and there are no screening inner-shell electrons, this poorly-shielded proton attracts lone pairs very strongly. The effect is so pronounced when hydrogen is bonded to fluorine, oxygen or nitrogen that the term 'hydrogen bond' is given to the bonds formed. For example, water molecules in ice:

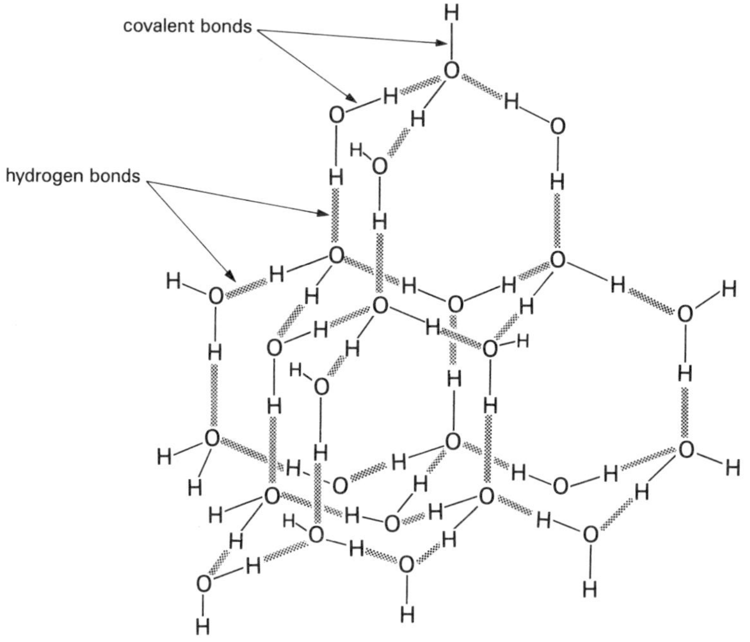

Molecular mass

The mass of an atom or a molecule is compared with the mass of an atom of carbon-12 as the chosen standard. This gives the relative atomic mass (A_r) or relative molecular mass (M_r) which are ratios and do not have units (they are 'pure' numbers).

> The relative molecular mass (M_r) of a substance is the weighted average of the masses of the molecules of the substance compared with one-twelfth of the mass of an atom of carbon-12.

The number of 'standard' carbon atoms in exactly 12 grams of carbon-12 is an important constant; it allows number relationships to be converted into mass relationships. This number is known as the Avogadro constant (L) and it follows, therefore, from the definition of relative molecular mass that L molecules of $M_r = m$ have a mass of m grams (the molar mass). The term *mole* is used as a measure of the amount of substance present in a sample containing L constituent particles. For example,

A mole of an atomic substance (e.g. diamond) is the amount of the substance that contains L atoms.
A mole of a molecular substance (e.g. water) is the amount of the substance that contains L molecules.

An unknown M_r is measured by using one of the following:

1 a mass spectrometer (for low values),
2 gas density measurements (see page 29).
3 data concerning the colligative properties of a solution (see page 40).

The mass spectrometer

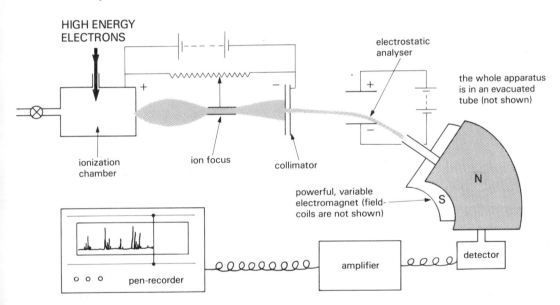

The mass spectrometer is an instrument which separates particles of different mass from one another. By calibrating the instrument with particles of known relative molecular mass, measurement of unknown relative molecular masses can then be carried out. The following functions are carried out in a mass spectrometer.

1 The sample is vaporized and introduced into the ionizing chamber.
2 The molecules lose electrons as a result of impact with the bombarding, high-energy electrons from the filament source.
3 A stream of cations is accelerated out of the chamber.
4 An electrostatic analyser allows only singly-charged cations to receive the correct deflection into the magnetic analyser.
5 The beam is deflected in the magnetic field of an electromagnet of variable field strength; for one particular value of field strength, only those cations of a particular mass receive the correct deflection to the detector.

6 The field strength is slowly and regularly increased, and the number of cations arriving at the detector is recorded.

There are two principal uses of mass spectrometry.

To calculate a precise value of A_r or M_r: a mass spectrometer trace at high sensitivity shows up the isotopic composition in a sample. For example, for gallium metal:

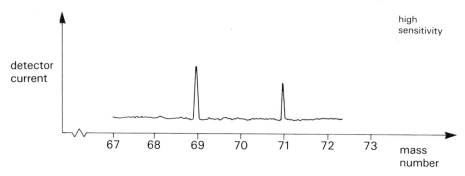

The height of each peak is proportional to the amount of each isotope: peak height at 69 = 1.5 cm; peak height at 71 = 1.0 cm.

Because A_r = the weighted average

$$A_r = \left[\frac{1.5}{1.5 + 1.0} \times 69\right] + \left[\frac{1.0}{1.5 + 1.0} \times 71\right]$$
$$= 69.8$$

To recognise an unknown substance: a mass spectrometer trace at low sensitivity gives a fragmentation pattern characteristic of the substance being analysed.

Molecular ions produced in the chamber break up into smaller fragments on their way to the detector. A typical trace shows small clusters of peaks at lower mass numbers than the molecular mass number, and these clusters provide a 'finger-print' of the particular substance. For example, the mass spectrograph of butan-1-ol (C_4H_9OH) is shown below.

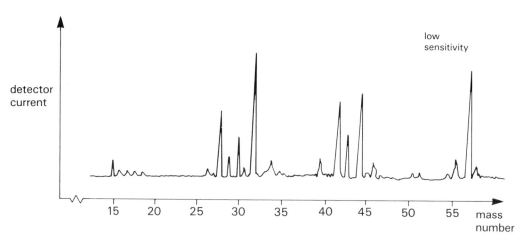

1.4 IONS

Metallic lattices

The most characteristic feature of a metal atom is its low degree of control of outer-shell electrons. This is a consequence of low core-charge, or large atomic radius, or both factors together.

In a metal lattice, all the atoms lose control of their outer-shell electrons to produce layers of cations in a cloud of delocalized electrons. These 'free' electrons are responsible for most of the characteristic properties of a metal.

The packing or arrangement of the cations in the lattice is described on page 25. The whole lattice is held together by 'metallic bonds' which are the forces of attraction between the cations and the delocalized electrons around them.

Ionic lattices

Although a metallic lattice contains cations, it would be incorrect to think of the lattice as ionic because of the presence of the delocalized electrons. An ionic lattice contains both cations and anions packed together.

By comparing the lattice energy and melting-point of a solid with similar data for other known solids, it is possible to estimate the percentage ionic character of the lattice. For example, consider the data for the three chlorides shown below.

chloride	lattice energy/kJ mol^{-1}	T_m/°C
NaCl	771	801
MgCl$_2$	2493	714
AlCl$_3$	5350	183 (sublimes)

The increased lattice energies reflect the decreasing radius of the cation concerned and its increasing charge. This is in agreement with the inverse square law which states that the force of attraction between two charged particles is directly proportional to the magnitude of the charges, and inversely proportional to the square of the distance between them.

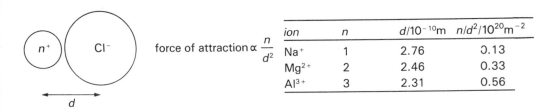

ion	n	$d/10^{-10}$m	$n/d^2/10^{20}$m^{-2}
Na$^+$	1	2.76	0.13
Mg^{2+}	2	2.46	0.33
Al^{3+}	3	2.31	0.56

The decrease in melting-point cannot be explained in terms of the simple ionic model. The decrease suggests a steady increase in the covalent character of the chlorides.

When aluminium chloride sublimes, the lattice breaks up into small molecular units rather than into isolated ions. Fajans' rules (page 16) suggest that the degree of covalency increases in the order NaCl, $MgCl_2$, $AlCl_3$ and the melting-point data appear to agree with this order.

In the diagram below, the sublimation of aluminium chloride is contrasted with the theoretical process taking place when the lattice energy is supplied.

Sublimation

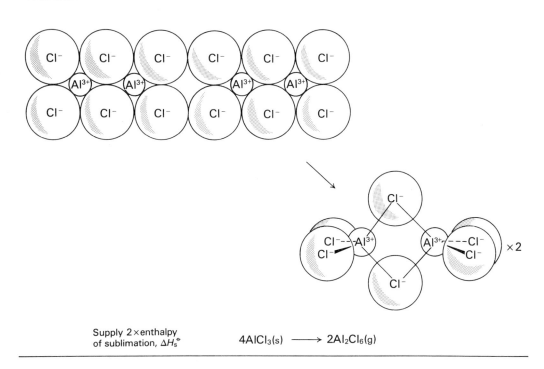

Supply 2×enthalpy of sublimation, $\Delta H_s^\ominus$

$4AlCl_3(s) \longrightarrow 2Al_2Cl_6(g)$

Lattice disruption

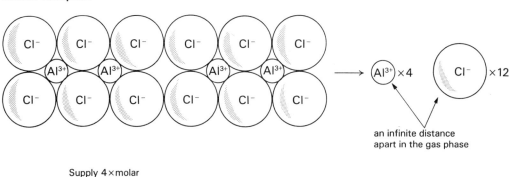

an infinite distance apart in the gas phase

Supply 4×molar lattice energy, U

$4AlCl_3(s) \longrightarrow 4Al^{3+}(g) + 12Cl^-(g)$

Packing arrangements

In a metal lattice, the cations are packed as closely as the forces of repulsion and attraction within the lattice allow. There are two main ways in which the cations are arranged.

1 Close-packed:
  (i) hexagonal close-packed (hcp),
  (ii) cubic close-packed (ccp), also called face-centred cubic (fcc):
2 Body-centred cubic packing (bcc).

Close-packed arrangements occupy 74% of the available space in the structure whereas body-centred cubic packing is more loose and occupies only 68% of available space. Each arrangement is illustrated below.

Close-packed

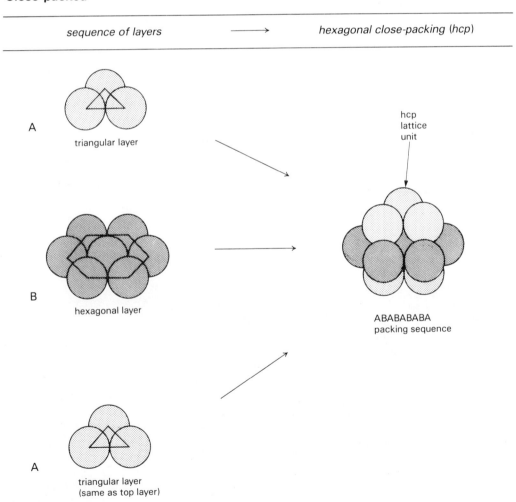

sequence of layers → hexagonal close-packing (hcp)

A — triangular layer

B — hexagonal layer

A — triangular layer (same as top layer)

hcp lattice unit

ABABABABA packing sequence

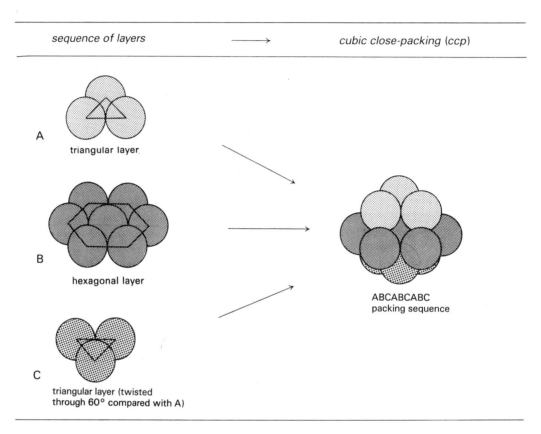

A cubic close-packed lattice gets its name from the fact that an ABCABC sequence of layers is also produced when 'face-centred' cubic units are packed together. Each cubic unit contains an ion at each corner and an ion at the centre of each face. A face-centred cube is shown below (with the ABC layers also illustrated).

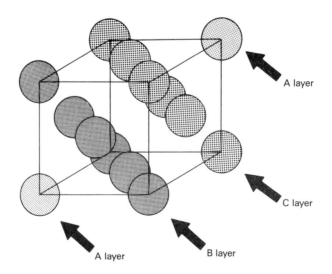

Body-centred cubic: this form of packing contains cubic units with an extra ion placed at the centre of each cube.

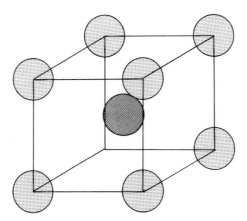

In an ionic lattice, both cations and anions are present. The packing arrangement, therefore, depends on:

1 the relative numbers of each ion in the lattice (the stoichiometry of the compound);
2 the relative sizes of each ion (the radius-ratio).

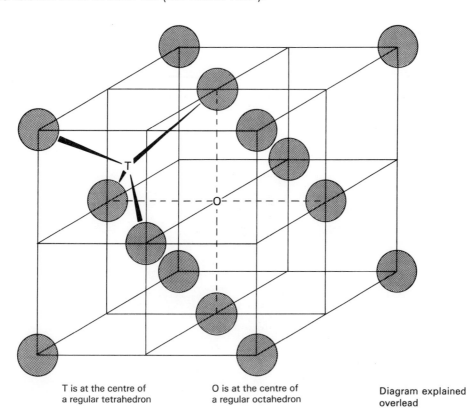

T is at the centre of a regular tetrahedron

O is at the centre of a regular octahedron

Diagram explained overlead

The packing in an ionic lattice may be described in terms of a close-packed arrangement of anions with the correct number of cations fitting in 'holes' between the anions. The proportion of the holes filled depends on the stoichiometry of the compound; whereas the choice of which holes to occupy in the lattice depends on the radius-ratio. For example, the diagram on the previous page shows a tetrahedral hole (T) and an octahedral hole (O) in a cubic close-packed lattice of anions.

When the cations are much smaller than the anions (radius-ratio $r^+/r^- \approx 0.3$), the tetrahedral holes are filled. Zinc sulphide forms a lattice of this sort. However, when the size of the cations increases until the radius-ratio exceeds 0.4, the cations are too big to fit into the tetrahedral holes. The octahedral sites are filled instead. Sodium chloride forms a lattice of this sort.

The major difference between the two arrangements is illustrated below.

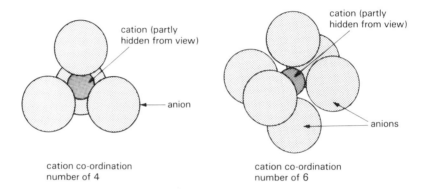

cation co-ordination number of 4

cation co-ordination number of 6

The number of near neighbours surrounding each ion is different. This number is known as the co-ordination number.

If the radius-ratio exceeds 0.73, even the octahedral holes are too small to accommodate the cations. The lattice opens up so that the co-ordination number increases to eight. Caesium chloride forms a lattice like this.

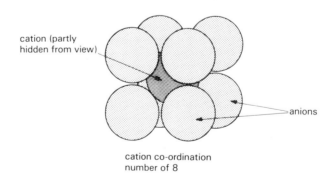

cation co-ordination number of 8

2 Phases

2.1 GAS

Gas density

The density of a substance in the gas phase is an important quantity because it can be used to provide information about the mass of the particles making it up. The relationship between gas density and particle mass is derived by applying Avogadro's hypothesis. Avogadro considered that the volume occupied by a single particle in the gas phase must be independent of the actual volume of the particle itself. The volume occupied is equivalent to the volume of space that the particle sweeps out, and this volume is a function only of the temperature and pressure. Hence the total volume occupied by a gas depends on the number of particles present.

> Avogadro's hypothesis states that, at the same temperature and pressure, equal volumes of two molecular gases contain equal numbers of molecules.

At 273 K and 1 atmosphere (0.1 MPa), a mole of any gas occupies the same volume (22.4 dm^3) because a mole of gas contains a fixed number of molecules (L, the Avogadro constant).

Hence, at the standard temperature and pressure (stp), the density of hydrogen is 2/22.4 g dm^{-3} while that of oxygen is 32/22.4 g dm^{-3}. The two densities are in the same ratio as the ratio of the two relative molecular masses. The relative molecular mass of any volatile substance can, therefore, be found by measuring its density in the gas phase. The following method is commonly used.

1. Two gas syringes are put empty into an oven whose temperature exceeds the boiling point of the unknown liquid.
2. A rubber sealing cap is placed over the end of each gas syringe.
3. Two micro liquid syringes are filled; one contains the unknown liquid, the other contains a volatile liquid of known M_r (for example, propanone).

4 The loaded liquid syringes are weighed, and then each is injected into a different gas syringe.
5 By weighing the discharged liquid syringes, the mass of injected liquid is calculated.
6 The volume taken up by the vapour at the oven temperature and pressure is read from the gas syringe scale in each case.
7 The densities ρ_1 and ρ_2 are then calculated and, from Avogadro's law, these are related to the two relative molecular masses, M_1 and M_2:

$$M_1/M_2 = \rho_1/\rho_2$$

For example, 0.136 g of propanone occupies 72 cm³ under the same conditions of temperature and pressure that 0.183 g of an unknown ester occupies 55 cm³.

$$\rho_1 = \frac{0.136}{72} \text{ g cm}^{-3} \quad \rho_2 = \frac{0.183}{55} \text{ g cm}^{-3} \quad M_1 = 58$$

$$\therefore M_2 = \left[58 \times \frac{0.183}{55} \times \frac{72}{0.136} \right] = 102$$

Kinetic theory

The pressure, volume and temperature of a gaseous system are related by two well-known experimental laws.

> Boyle's law states that, at constant temperature, the volume of a fixed mass of gas is inversely proportional to its pressure; Charles' law states that, at constant pressure, the volume of a fixed mass of gas is directly proportional to its temperature.

The kinetic theory of gases developed out of the attempts to explain these relationships in terms of the behaviour of particles in the gas phase. Five main assumptions are made.

1 Molecules in the gas phase are hard spheres in random motion.
2 There are no forces of attraction between the molecules.
3 The collisions between the molecules and the container wall and between the molecules themselves are elastic.
4 The duration of a collision is negligibly short compared with the time interval between collisions.
5 The volume occupied by the molecules themselves is negligibly small compared with the volume of the container.

These assumptions are used in the following way to interpret the three system functions, pressure, volume and temperature.

Pressure. When a molecule collides with the container wall, it exerts a force because there is a change in momentum as a result of the collision. The pressure exerted by a gas is equal to the sum of these forces divided by the surface area of the container wall.

At a constant temperature, the frequency of the collision determines the pressure exerted by a gas. In a mixture of gases, it is, therefore, possible to express the total pressure exerted in terms of the relative numbers of each type of particle present. By defining the *partial pressure* of a component in the mixture as shown below, the total pressure is given as the sum of all the partial pressures. This is known as Dalton's law.

> The partial pressure, p_A, exerted by a component, A, in a gas mixture is equal to its mole fraction, x_A, times the total pressure, p_T. $p_A = x_A \times p_T$.

It therefore follows that the partial pressure of a component is equal to the pressure that would be exerted if it alone occupied the container (i.e. $x_A = 1$).

Volume. The volume occupied by a gas is imposed only by the dimensions of the container. An uncontained gas has an infinite volume.

Temperature. There is a direct correlation between the kinetic energy of the molecules and the temperature of the system. Temperature is, therefore, seen as being a measure of the mean kinetic energy of the molecules ($\overline{ke} = \frac{1}{2}m\overline{c^2}$ where m is the mass of a molecule and $\overline{c^2}$ is the average of the squares of the molecular speeds.)

$$T \propto \tfrac{1}{2}m\overline{c^2}$$

As a result of the collisions within the system, the molecules are unlikely to be travelling at a constant speed. The distribution of molecular speeds follows a form of the Maxwell–Boltzmann function and is shown below for two temperatures.

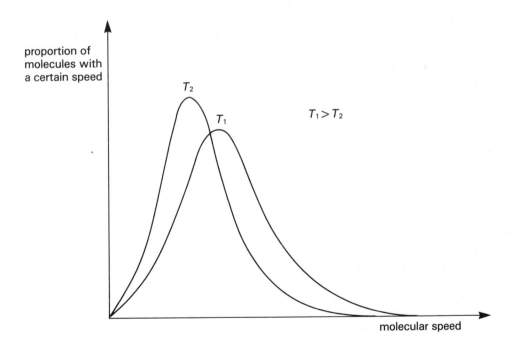

p, V and T dependence. By calculating the rate of change of momentum at the walls of a cubic container of volume V, it can be shown that the pressure p depends on the number of molecules N, their mass m and their mean squared speed $\overline{c^2}$.

$$pV = \tfrac{1}{3}Nm\overline{c^2}$$

But, since $T \propto \tfrac{1}{2}m\overline{c^2}$, then $pV \propto NT$.

In other words:

1 for a fixed mass of gas at constant temperature, pV = a constant—Boyle's law;
2 for a fixed mass of gas at constant pressure, $V \propto T$—Charles' law;
3 at constant pressure and temperature, $V \propto N$—Avogadro's hypothesis.

Ideal gas

An ideal gas is a gas whose behaviour fits the assumptions of the kinetic theory. The pressure, volume and temperature of an ideal gas are related by the derived equation $pV \propto NT$. For n moles of gas, this ideal gas equation is given in the following form:

$$pV = nRT$$

R is called the ideal gas constant and has a value of 8.314 J K^{-1} mol^{-1} when p is measured in N m^{-2} (Pa), V in m^3 and T in kelvins.

A real gas deviates from ideality in two main ways.

1 The volume occupied by the molecules themselves is *not* negligible, particularly at high gas pressures. A 'real' gas of volume V, has an 'ideal' value only when a small quantity b, is subtracted from it, i.e. ideal volume = $(V - b)$.
2 There *are* forces of attraction between the molecules. At high pressure or at low temperature, these forces have their greatest effect. A molecule about to collide with the container wall experiences a retarding force due to its attraction to those in the bulk of the gas. When the retarding force is summed for all the molecules, the reduction in pressure depends inversely on V^2. So a 'real' gas of pressure p, has an 'ideal' value only when a small quantity a/V^2 is added to it, i.e. ideal pressure = $(p + a/V^2)$.

By including these two corrections in the ideal gas law, van der Waals' equation of state is obtained.

$$(p + a/V^2)(V - b) = nRT$$

a and b are constants whose values are different for different gases.

The pressure dependence of the product pV illustrates the effects of the two deviations from ideality described above. pV is plotted against p both for an ideal gas and for nitrogen on the following graph.

1 At low pressure, $(pV)_{real} < (pV)_{ideal}$ because $p_{real} < p_{ideal}$ as a result of molecular attraction.
2 At high pressure, the above factor is overcome because the molecular volume becomes increasingly significant as the space between the molecules is reduced: $V_{real} \geqslant V_{ideal}$ at high pressure and so $(pV)_{real} > (pV)_{ideal}$.

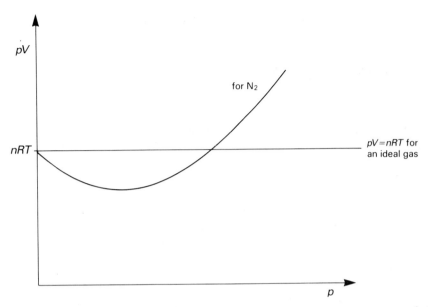

The pressure dependence of the volume of a gas at a constant temperature is known as an isotherm. Andrews' work on the isotherms of carbon dioxide gave curves which agreed quite well with the van der Waals' equation of state.

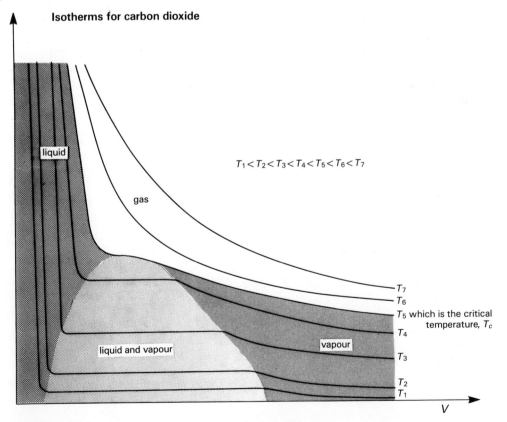

Above the critical temperature T_c, the isotherms show no discontinuities or inflection points because the gas cannot be liquefied no matter how much pressure is exerted. Below T_c, at a particular value of pressure, the volume suddenly decreases sharply as liquefaction occurs. Subsequent increases in pressure then cause almost no reduction in volume because liquids are almost incompressible. It is usual to describe the gas phase of a substance below its critical temperature as a vapour.

2.2 LIQUID

Vapour pressure

Although the particles in a liquid are packed nearly as tightly as those in a solid, there is no long-range order or 'liquid' lattice. A particle tends to vibrate about its mean position in a 'clump' of other particles which may possess short-range order. These clumps move slowly and randomly through the system, frequently altering their boundaries as particles move from clump to clump.

The distribution of energy amongst the particles resembles that of the molecular speeds in gases (see page 31). The significant differences are that the range of energies is much smaller, and that the vast majority of particles have energies close to the mean. The few that have higher energy at the surface of the liquid have a tendency to escape from the system. These escaping particles are responsible for the process called evaporation. A liquid, therefore, exerts a vapour pressure as a result of the escaping tendency of its particles.

As the temperature increases, so does the mean kinetic energy of the particles and, therefore, their escaping tendency also increases. If the particles in a liquid have a high escaping tendency, the liquid exerts a high vapour pressure and boils at a low temperature. The boiling-point T_b, is the temperature at which the liquid's vapour pressure equals the external pressure.

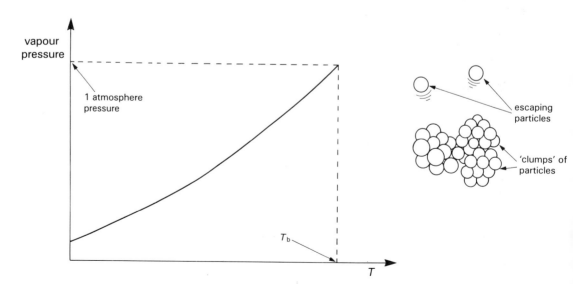

Raoult's law

In a solution (a one phase mixture), the escaping tendency of a particular component's particles depends on the probability of those particles being at the surface. Providing that all the forces between the particles are the same no matter what the particles are, the following logic can be applied:

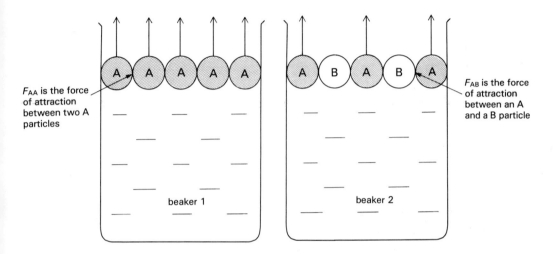

The escaping tendency of Ⓐ particles from beaker 2 is 60% of that from beaker 1.

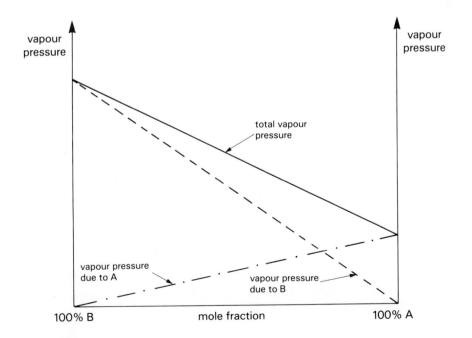

Raoult's law states that, for an ideal solution at constant temperature, the vapour pressure exerted by a given component equals its mole fraction times the vapour pressure of the component when pure. An ideal solution is one in which all the forces between the particles are constant: $F_{AA} = F_{AB} = F_{BB}$.

Volatile solutes are solutes whose vapour pressure is of the same order as that of the solvent. The total vapour pressure exerted by the solution is given by Dalton's law (page 31), and this is illustrated on the graph (page 35). Since solute and solvent are unlikely to have the same escaping tendency, the composition of vapour evaporating from a solution is unlikely to be equal to the composition of the liquid. The vapour will be richer in the component with the higher escaping tendency.

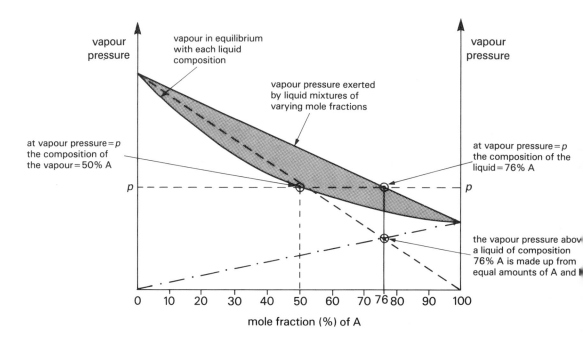

A solution of composition 76% A has vapour above it containing only 50% A: the escaping tendency of B is greater than the escaping tendency of A.

Non-volatile solutes exert no vapour pressure. Their particles simply block the solvent particles at the solution surface. Raoult's law applies differently here because both the solvent and the total vapour pressure are lowered.

The relative lowering of the vapour pressure of a pure solvent at constant temperature is equal to the mole fraction of non-volatile solute added.

This alternative form of Raoult's law can be deduced as follows. Given that p_0 is the vapour pressure of pure solvent, p is the vapour pressure of the solution, n_0 is the number of moles of solvent and n the number of moles of solute present:

1 apply Raoult's law to find the solvent vapour pressure, p: $p = \underbrace{[n_0/(n_0 + n)]}_{\text{mole fraction of solvent}} \times p_0$

2 the relative lowering of the solvent vapour pressure is: $(p_0 - p)/p_0$

From 1 the expression for p can be substituted into 2.

the relative lowering of the solvent vapour pressure
$$(p_0 - p)/p_0 = [p_0 - [n_0/(n_0 + n)] \times p_0]/p_0$$
$$= p_0[1 - n_0/(n_0 + n)]/p_0$$
$$= 1 - n_0/(n_0 + n)$$
$$= [(n_0 + n) - n_0]/(n_0 + n) = \underbrace{n/(n_0 + n)}_{\text{the mole fraction of solute present}}$$

Azeotropes

In the same way that the behaviour of real gases deviates from ideality, the properties of real solutions differ from those of an ideal solution. The differences are a consequence of the inequality of the forces between the particles of a solution. If $F_{AB} > F_{AA}$ and F_{BB}, it becomes more difficult for an A particle to escape when surrounded by B particles. Similarly, it is more difficult for B particles to escape from a cluster of A particles. If the deviations are large enough, a minimum is produced in the total vapour pressure curve. The diagram below should be compared with that for an ideal solution of page 35.

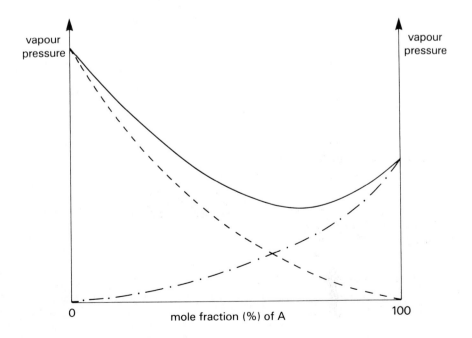

38 Condensed Chemistry

The effect is described as a negative deviation from Raoult's law because the vapour pressure is lower than the value predicted by the law.

If $F_{AB} < F_{AA}$ or F_{BB}, the opposite of the above paragraph holds true. The effect is then described as a positive deviation from Raoult's law because the vapour pressure is greater than the value predicted by the law.

Distillation. The two deviations are often shown as functions of boiling point rather than as functions of vapour pressure. Since the boiling point of a solution is inversely dependent on its vapour pressure, the vapour pressure curves become inverted when shown as boiling point curves. In the diagrams on the facing page, the vapour composition curves are included as well. These are also inverted compared with their position on the vapour pressure curve on page 36.

If a liquid of composition L_1 is boiled, it boils at temperature T_1. The vapour above it is richer in B, the component with the lower boiling point; the composition of the vapour at this temperature is V_1. When V_1 is liquefied and boiled again, a second vapour V_2 is

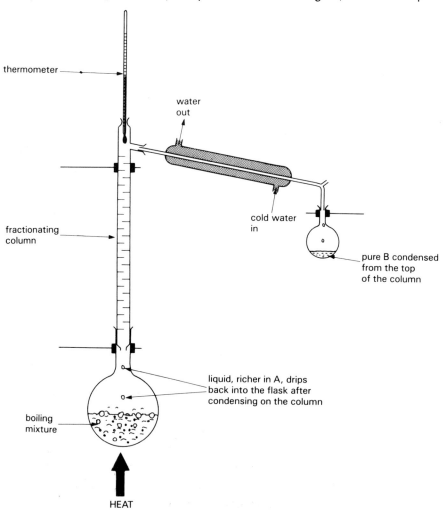

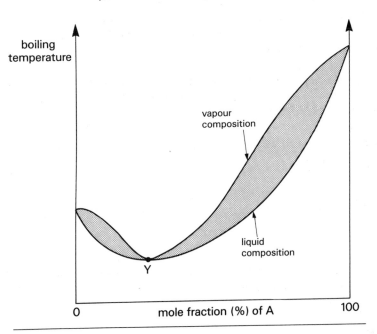
positive deviation (e.g. ethanol/water)

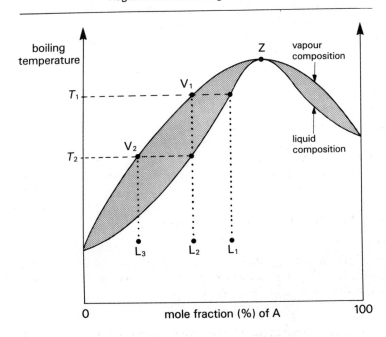
negative deviation (e.g. nitric acid/water)

produced, richer again in B. By repeating this process many times, pure B can be driven off as a vapour while the liquid becomes gradually richer in A.

In the fractionating column shown on page 38, a sequence of successive liquefaction and boiling takes place. The liquid in the flask gradually becomes richer in A as pure B is driven from the top of the column.

When the liquid composition reaches Z, the boiling temperature reaches its maximum and the vapour has the same composition as the boiling liquid. A solution with these properties is known as an azeotrope.

Similar conclusions can be drawn about the distillation of a solution with a positive deviation from Raoult's law. However, there are two important differences.

1 The *vapour* becomes azeotropic, and the liquid in the flask gradually becomes pure A or B.
2 The azeotrope, Y is a minimum boiling-point mixture rather than a maximum boiling-point mixture.

> An azeotrope (or constant boiling mixture) is a solution whose composition remains unchanged when the solution is boiled. The vapour above the solution has the same composition as that of the liquid.

Colligative properties

When a non-volatile solute is added to a solvent, the vapour pressure decreases because the mole fraction of the volatile component decreases. Since the number of non-volatile solute particles per unit volume controls the drop in vapour pressure, any property related to this drop in vapour pressure is called a colligative property ('colligative' means 'depending on concentration').

There are three main colligative properties:

1 elevation of boiling point,
2 depression of freezing point,
3 the osmotic pressure of a solution.

Elevation of boiling point. By Raoult's law, the relative lowering of the solvent vapour pressure equals the mole fraction of the non-volatile solute present. Because the vapour pressure of a solution is inversely proportional to its boiling point, a relative lowering of vapour pressure brings about an elevation in boiling point. This elevation is, therefore, proportional to the mole fraction of solute present.

> The ebullioscopic (or boiling point) constant for a given solvent is the elevation in boiling point caused by the addition of a mole of solute to 1000 grams of solvent at a pressure of 1 atmosphere (0.1 MPa).

When the ebullioscopic constant is known for a solvent, it becomes possible to measure M_r for an unknown non-volatile solute: a known mass of solute is added to a known mass of solvent and the elevation in boiling point is determined. For example, given $K_b = 1.86$ K mol^{-1} in 1000 g for water, and that 2.00 g of a white solid in 100 g water gave a solution boiling at a temperature 0.21 K higher than the boiling temperature of pure water M_r is found as follows:

  2.00 g in 100 g of water ⟶ $\Delta T = 0.21$ K (given)

  ∴ 20.00 g in 1000 g of water ⟶ $\Delta T = 0.21$ K

  ∴ $(20.00 \times 1.86/0.21)$g in 1000 g ⟶ $\Delta T = 1.86$ K

  ⇒ 177 g in 1000 g of water ⟶ $\Delta T = 1.86$ K

  ∴ M_r of white solid is 177.

Depression of freezing point. The freezing point of a liquid depends directly on the vapour pressure that the liquid and solid exert. A relative lowering of vapour pressure, therefore, brings about a proportional depression in freezing point.

> The cryoscopic (or freezing point) constant for a given solvent is the depression in freezing point caused by the addition of a mole of solute to 100 grams of solvent at a pressure of 1 atmosphere (0.1 MPa).

When the cryoscopic constant is known for a solvent, the measurement of a depression in freezing point can also be used to determine M_r for a solute.

Osmotic pressure. If a solution is separated from pure solvent by a semipermeable membrane, solvent flows through the membrane into the solution.

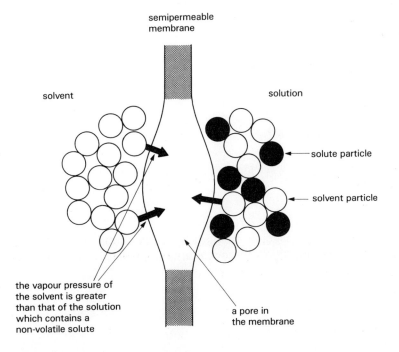

The flow of solvent is caused by the inequality of the vapour pressure on the two sides of the membrane. The process is called osmosis, and the pressure required to prevent osmosis from taking place is called the osmotic pressure of the solution.

> The osmotic pressure of a solution, when separated from its pure solvent by a semipermeable membrane, is the pressure that must be applied to the solution to prevent the inflow of solvent.

In a solution that contains n moles of solute, the osmotic pressure Π, is inversely proportional to the volume V, of the solution and is directly proportional to the absolute temperature T. By analogy with the ideal gas law,

$$\Pi V = nRT$$

One of the best methods of determining M_r for an unknown polymer makes use of the above relationship. The osmotic pressure of a dilute solution of known concentration is measured. Even though this pressure is often of the order of a few millimetres of mercury (~500 Pa), the method is reliable for relative molecular masses as high as 100 000 or so.

For example, 2.41 g of a polymer dissolved in 1000 cm³ (10^{-3} m³) of water causes an osmotic pressure at 15°C of 1.7×10^3 Pa. Calculate M_r for the polymer.

Let there be n moles of polymer present in 2.41 g

Since $n = \dfrac{\Pi V}{RT}$ where $\Pi = 1.7 \times 10^3$ N m^{-2}

$$V = 10^{-3} \text{ m}^3$$

$$R = 8.31 \text{ J mol}^{-1} \text{ K}^{-1}$$

and $T = 298$ K

$$\therefore n = \frac{1.7}{8.31 \times 298} = 6.86 \times 10^{-4}$$

∴ in one mole, there are $(2.41/n)$g = 3510 g

∴ $M_r = 3510$.

The ebullioscopic or cryoscopic methods are less useful for determining M_r for a polymer because the temperature differences are too small to be measured with accuracy.

2.3 SOLID

Lattice type

A lattice is classified according to the type of particle that is fundamental to its structure.

1. Atomic lattices: (i) metallic; (ii) macromolecular.
2. Molecular lattices.
3. Ionic lattices.

Atomic lattices. These include both metallic and macromolecular structures. The two differ only in the arrangement of the outer-shell electrons of the atoms in the lattice.

lattice structure	outer-shell electrons	properties of solid
(i) metallic e.g. sodium 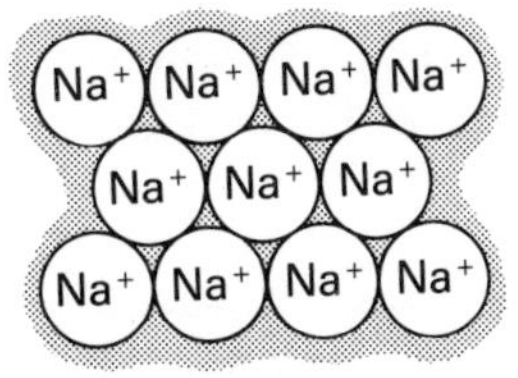	the outer-shell electrons are delocalized throughout the whole lattice.	malleable; ductile; good thermal and electrical conductors; range of hardnesses and fixed points
(ii) macromolecular e.g. silicon(IV) oxide 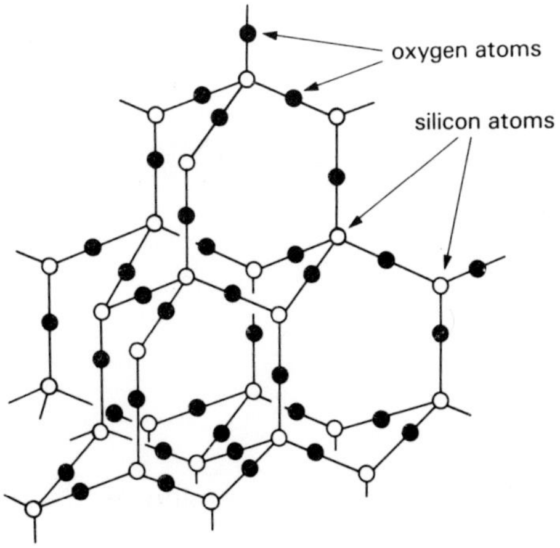	the outer-shell electrons are localized in pairs between the atoms throughout the lattice	very hard; thermal and electrical insulators; insoluble in all solvents; high fixed points

Molecular lattices. These contain discrete molecules within the lattice. Each molecule is attracted to its neighbours by van der Waals' forces.

lattice structure	outer-shell electrons	properties of solid
molecular e.g. carbon dioxide	the outer-shell electrons are localized in pairs within each molecule	weak; soft; thermal and electrical insulators; soluble in non-polar solvents; low fixed points

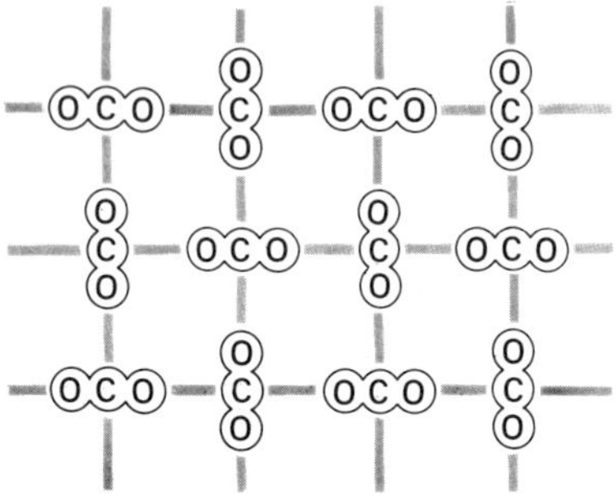

Ionic lattices. These contain anions and cations within the lattice. Electrostatic forces of attraction between the ions hold the whole lattice together.

lattice structure	outer-shell electrons	properties of solid
ionic e.g. sodium chloride	the outer-shell electrons are localized within each anion; the cations have lost their outer-shell electrons	hard but brittle; electrical conductors in the molten state or in aqueous solution; high fixed points

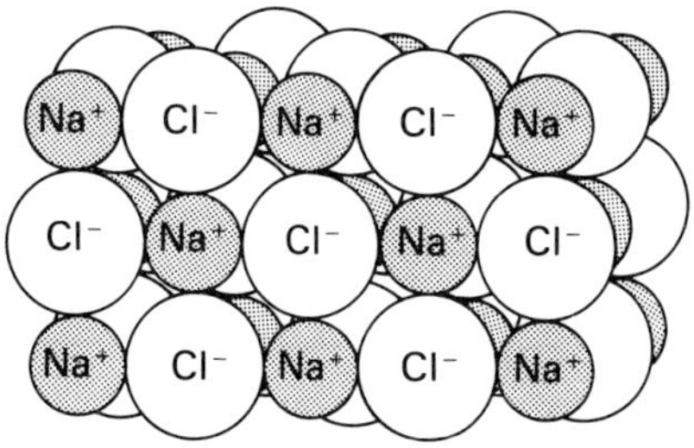

Determination of a lattice structure

A lattice contains various different planes of particles extending throughout its structure. For example, in the simplified, two-dimensional array shown on the facing page, two sets of parallel planes are picked out.

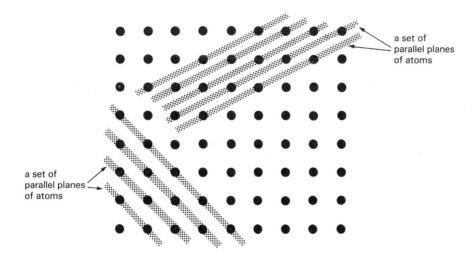

The distance between consecutive parallel planes is of the same order as the wavelength of X-rays (electromagnetic radiation, $\lambda \approx 10^{-10}$ m). When a beam of monochromatic X-rays strikes a crystal, a diffraction pattern can be observed due to the interference of waves reflected from successive parallel planes. When X-rays are reflected from a particular set of planes at an angle, θ, destructive interference cancels out the reflection unless the following condition holds (the Bragg equation):

$n\lambda = 2d \sin \theta$ $\quad d$ = the spacing between the planes
$\quad\quad\quad\quad\quad\quad\quad\lambda$ = the wavelength of the X-rays
$\quad\quad\quad\quad\quad\quad\quad n$ = any whole number

Bragg derived this equation as follows.

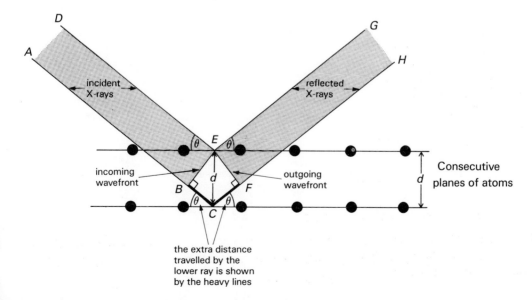

The extra distance travelled by the wave hitting the lower plane $= (BC + CF)$.
But in triangle BEC, $BC = d \sin \theta$ and in triangle CEF, $CF = d \sin \theta$

∴ extra distance travelled $= 2d \sin \theta$

For constructive interference of the reflected wave, both FH and EG must be in phase. In other words, the extra distance must be a whole number of wavelengths:

∴ $n\lambda = 2d \sin \theta$ where $n = 1, 2, 3 \ldots$

The Bragg equation is used to analyse the diffraction pattern produced when a crystal is irradiated in a special X-ray camera. An intense monochromatic beam of X-rays is obtained by accelerating high energy electrons (~50 kV) at a copper anode. With suitable filters and collimators, a parallel beam of wavelength $= 1.542 \times 10^{-10}$ m is generated. The beam is shone at a crystal surrounded by photographic paper in the camera unit. The diffraction pattern appears as a number of exposures at different angles from the incident beam. With a knowledge of the dimensions of the camera, θ can be established for each exposure and since λ is known, the symmetry and spacing of the planes can be deduced.

Eutectics

When a few crystals of a substance dissolve in the molten phase of another substance, the melting point of the second substance is depressed (see page 41). Exactly the same effect occurs if crystals of the second substance are added to a pure melt of the first substance. At one particular composition, therefore, a minimum in melting point is reached. The minimum represents the meeting of the two depression trends. This composition is known as a eutectic.

A eutectic is a mixture that melts at a single fixed temperature like a pure substance: the composition of the melt equals that of the solid.

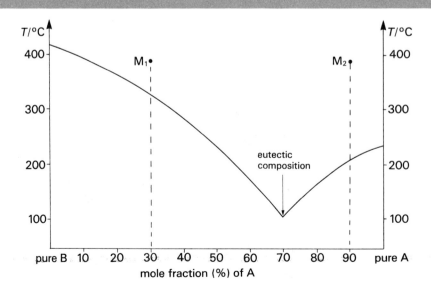

The lattice structure of a eutectic has a regular pattern in which the particles of the different components are present in fixed proportion.

Thermal analysis. The composition of a eutectic is determined by a method called thermal analysis.

1. Mixtures of different composition are made up by weighing out different amounts of each component.
2. The mixtures are melted and, on cooling, their temperatures are recorded over a period of time.
3. Cooling-curves are plotted for each mixture of known composition.
4. A pure substance (or eutectic) has a cooling-curve with a single inflection at its phase-change temperature.
5 Other mixtures have a cooling-curve with two inflections: the first occurs at the temperature at which pure A or B starts to solidify; the second occurs when the melt reaches the eutectic composition.

For example, compositions M_1 and M_2 shown on the diagram on page 46 have the following cooling-curves. Each has two inflection points:

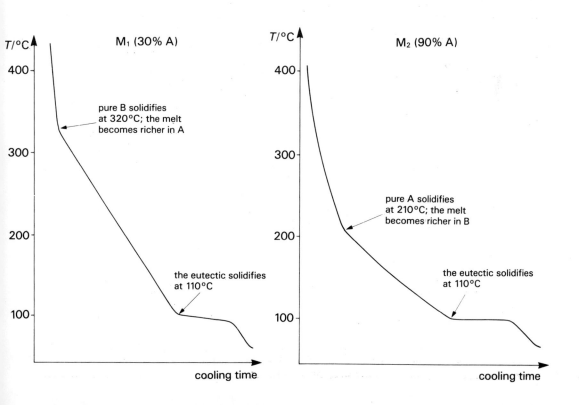

When all the range of different melts have been cooled, the phase change data is collected on a single graph. This gives the phase diagram shown below.

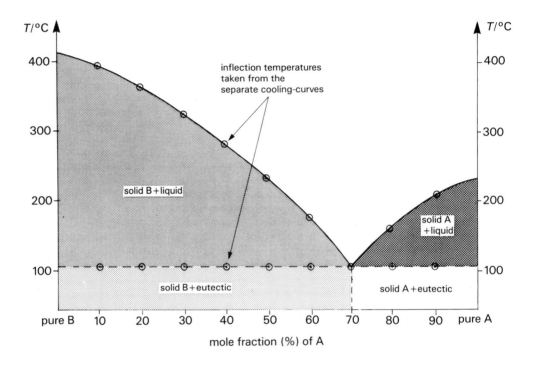

3 Physical Equilibrium

3.1 PHASE CHANGES IN PURE SUBSTANCES

Phase diagrams

The phase diagram of a substance shows the conditions of temperature and pressure required for the stability of each phase. For example:

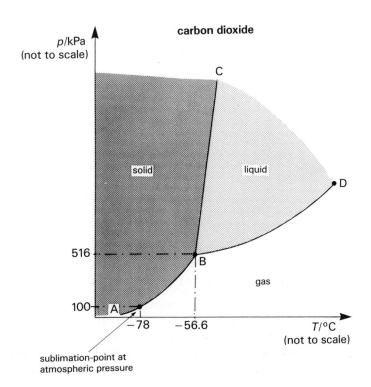

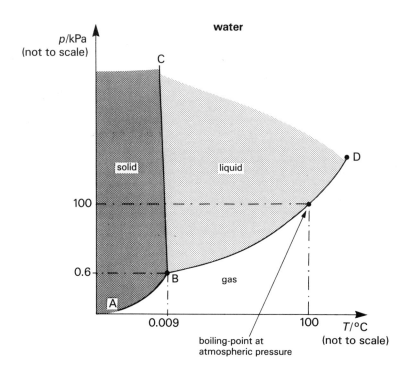

The lines on the phase diagram represent the conditions of temperature and pressure under which one phase changes into another. Under these conditions, however, the reverse process also takes place: the second phase is changing back into the first. The rates of the two opposing processes are equal and the system is, therefore, in physical equilibrium. The system remains unaltered unless the rate of one of the two opposing processes changes, and this only occurs if the temperature or pressure is altered.

line or point	conditions of temperature and pressure under which the following phases are in equilibrium
AB	solid and gas: sublimation points at different pressures $(s) \rightleftharpoons (g)$
BC	solid and liquid: melting points at different pressures $(s) \rightleftharpoons (l)$
BD	liquid and gas: boiling points at different pressures $(l) \rightleftharpoons (g)$
B	solid, liquid and gas: the triple point $(s) \rightleftharpoons (l) \rightleftharpoons (g)$

There are three other significant features of the two phase diagrams shown above.

1. Point D is known as the critical temperature. Above this temperature it is impossible to liquefy a gas by increasing the pressure.
2. Solid carbon dioxide sublimes at atmospheric pressure when the temperature rises above −78°C: the pressure at the triple point is well above atmospheric pressure.
3. The solid-liquid phase line, BC slopes to the right for most substances except water.

An increase in pressure at constant temperature usually causes a liquid to solidify as the particles are pressed closer together. However, since ice is less dense than water, the opposite is true for water: ice can be melted by increasing the external pressure on it at a constant temperature. Water molecules pack less tightly in the solid phase than in the liquid phase because of the extensive ordering effect caused by hydrogen bonding in ice. The structure of ice is shown on page 20. Each water molecule has a co-ordination number of four, whereas in liquid water, some molecules have five near neighbours because the hydrogen bonding is less well ordered.

Allotropy

Allotropy is the property of an element to exist in more than one physical form in the same phase. The forms differ in their structures or in the packing arrangements of their lattice. There are three major types of allotropy.

Enantiotropy. One allotrope is stable under one set of conditions of temperature and pressure; another allotrope is stable under a different set of conditions. For example, sulphur exhibits enantiotropy:

The lines and areas on the above phase diagram are interpreted in the same way as before. For example,

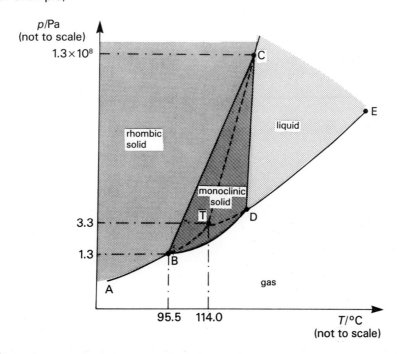

BC represents the conditions of temperature and pressure under which the two solid enantiotropes are in equilibrium: rhombic ⇌ monoclinic.
T represents the hypothetical triple point and BT, TC and TD are metastable phase lines. Metastability is discussed on page 53.

Monotropy. One allotrope is always the more stable no matter what the conditions of temperature and pressure are. For example, phosphorus exhibits monotropy:

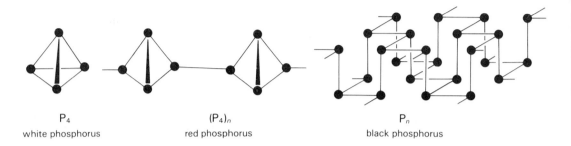

P_4
white phosphorus

$(P_4)_n$
red phosphorus

P_n
black phosphorus

The two common allotropes are white and red. Black phosphorus can be obtained by heating red phosphorus to a high temperature under very high pressure.

The monotropic relationship of red and white phosphorus is illustrated by the variation of vapour pressure with temperature shown below.

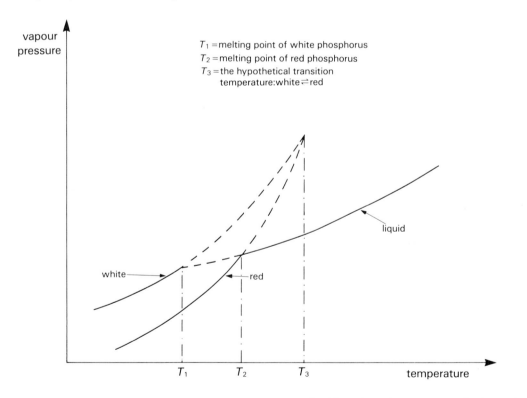

At any single temperature, the vapour pressure of white phosphorus exceeds that of red because the escaping tendency of P_4 molecules is lower from the chain structure of red phosphorus. The hypothetical transition temperature, T_3 between the two allotropes, is the temperature at which metastable white and metastable red phosphorus are in equilibrium. However, since T_3 is well above the melting point of both allotropes, it is impossible for them to reach equilibrium. Red is always more stable.

Dynamic allotropy. This is exhibited by an element existing in more than one molecular form in the same phase. These forms are in dynamic equilibrium with one another. For example, in sulphur vapour:

$$S_8 \rightleftharpoons 2\, S_4 \rightleftharpoons 4\, S=S$$

Metastability

During a phase change, the arrangement of the particles in the system alters. Although the new arrangement is of lower potential energy, an activation energy (page 73) must be supplied in order to bring about the change. This energy factor is sometimes quite considerable. For example, graphite is the more stable monotrope of carbon, and yet diamonds do not readily change into graphite, even though the overall process is energetically favourable. The activation energy needed to rearrange a diamond lattice is vast. Diamond is, therefore, a *metastable* phase of carbon.

> A metastable phase exists at a set of conditions under which a different phase is the more stable.

The rate at which a metastable phase changes into the more stable phase is often imperceptibly slow.

Some different aspects of metastability are illustrated by:

1 supercooling,
2 phase changes in sulphur.

Supercooling. The following cooling curve is typical for a pure liquid near its freezing point.

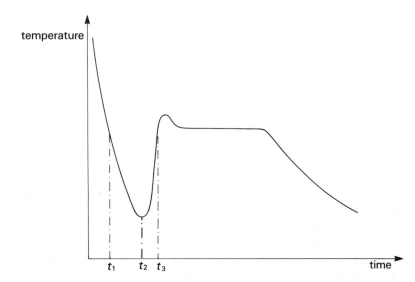

At time t_1, the temperature of the melt reaches the freezing point. No phase change takes place because, during the rapid cooling, the necessary activation energy for the formation of the solid lattice has not been supplied. At time t_2, a small section of solid lattice starts to form and, as the reorganization spreads rapidly through the system, energy is given out because new bonds are made. The temperature increases sharply as this energy is evolved, and some melting takes place as a result. Finally, the temperature falls to the freezing point again but now, in the presence of some seed crystals of solid, the whole system changes phase. Between times t_1 and t_3, the liquid is metastable and is said to be supercooled.

Phase changes in sulphur. When rhombic sulphur is heated at atmospheric pressure, monoclinic sulphur should be produced at the temperature T_1 represented below.

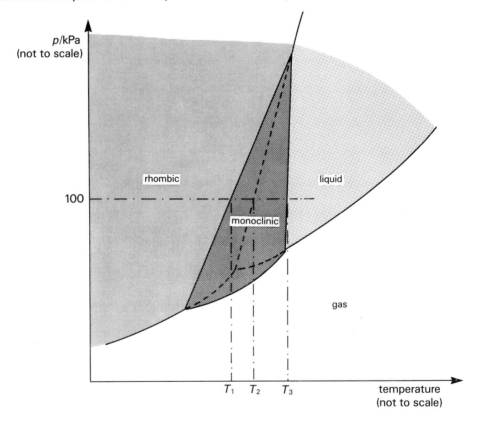

In practice, this does not take place because the activation energy for the process is too great. Instead, metastable rhombic sulphur is produced and this melts at T_2 to give metastable liquid sulphur. The liquid is metastable until it reaches a temperature of T_3: between T_1 and T_3 the most stable phase is monoclinic sulphur. By cooling liquid sulphur very slowly through a temperature range close to T_3, monoclinic sulphur can be obtained. However, if liquid sulphur is cooled rapidly, it becomes metastable at T_3 and solidifies at T_2 to give metastable rhombic sulphur again. Liquid sulphur is discussed in greater detail on page 136.

3.2 PHASE CHANGES IN SOLUTIONS

The sections on azeotropes (page 37) and eutectics (page 46) outline the characteristic ways in which a solution boils or freezes. Like the phase changes that occur in pure substances, those that occur in mixtures are also governed by the principles of physical equilibrium. There are two further aspects of phase equilibria in solutions to be discussed: solubility and partition.

Solubility

When a solute is added to a solvent, the forces of attraction between the particles of the two separate substances must be overcome before they can mix. The energy required for this process is often supplied by the energy released as new forces of attraction are set up between solute and solvent particles. The process of mixing is favoured by the increase in the degree of disorder of the system.

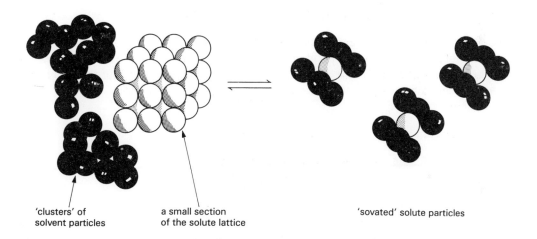

'clusters' of solvent particles

a small section of the solute lattice

'sovated' solute particles

As soon as there are some solvated solute particles present in the system, the reverse process can also start to take place: pure solute crystallizes if the solute particles break free from solvent particles and reform a lattice. When the rates of the two opposing processes are equal, the system is in equilibrium and no more solute can dissolve. The solution is then said to be saturated (providing there is excess solute in contact with it).

Solubility product K_{sp}. In an aqueous solution of ions at a given temperature, the rate of formation of the solid lattice depends on the concentration of the ions present. However, the rate of the opposing process (the rate of dissolution) depends only on the concentration of the solid; a constant. Hence, the equilibrium condition depends only on the concentration of dissolved solute present in a saturated solution.

This condition is described by an equilibrium constant (see page 80), called the solubility product K_{sp}. For an ionic solute M_aX_b dissolving in water to produce a saturated solution, the equilibrium concentration (in mol dm^{-3}) of the ions are related as shown overleaf.

$$M_aX_b(s) \rightleftharpoons aM^{b+}(aq) + bX^{a-}(aq)$$

At equilibrium,

$$K_{sp} = [M^{b+}]^a[X^{a-}]^b \quad \text{where [ ] means 'molar concentration of'}$$

For example, calcium hydroxide $Ca(OH)_2$ has $a = 1$ and $b = 2$,

$$K_{sp} = [Ca^{2+}][OH^-]^2 = 5.5 \times 10^{-6} \text{ mol}^3 \text{ dm}^{-9} \text{ at 298 K}$$

The value of the solubility product is a measure of the amount of solute that can be dissolved in a particular volume of solvent. For example, its value can be used to predict whether precipitation of calcium hydroxide is likely or not when a sodium hydroxide solution is added to one of calcium chloride. Given that the calcium chloride contains 2.2 g dm^{-3} and that an equal volume of 0.2 mol dm^{-3} sodium hydroxide is added, the ionic concentrations in the mixture are:

$$[Ca^{2+}] = (\tfrac{1}{2} \times 2.2/111) = 1 \times 10^{-2} \text{ mol dm}^{-3}$$
$$[OH^-] = (\tfrac{1}{2} \times 0.2) = 1 \times 10^{-1} \text{ mol dm}^{-3}$$

The factor of a half is necessary because of the dilution effect when equal volumes of the two solutions are mixed. The ionic product $[Ca^{2+}][OH^-]^2$ exceeds the value of the solubility product and, therefore, calcium hydroxide is precipitated until the ionic concentrations fall sufficiently to allow the equilibrium conditions to apply.

On mixing,

$$[Ca^{2+}][OH^-]^2 = 1 \times 10^{-2} \times (1 \times 10^{-1})^2 = 1 \times 10^{-4} \text{ mol}^3 \text{ dm}^{-9}$$

Precipitation of calcium hydroxide must occur until:

$[Ca^{2+}][OH^-]^2$ has fallen to 5.5×10^{-6} mol^3 dm^{-9} (K_{sp})

Distribution

Solutes are often soluble in several different solvents. If a solute dissolves in two immiscible solvents present in the same beaker, transfer of solute takes place across the phase boundary in both directions. The rate of transfer varies with the solubility of the solute in each solvent and with the concentrations present in each layer. When the two rates are equal, the system is in equilibrium.

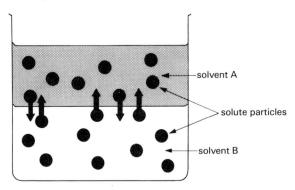

The ratio of the solute concentrations in the two solvents is constant at constant temperature when the system is in equilibrium. This constant is called the distribution constant, K, for the two solvents and the solute concerned.

Solvent extraction. Distribution is used in a separation technique called solvent extraction.

For example, phenylamine is far more soluble in chloroform than in water. In the preparation of phenylamine, it is produced as an emulsion in water. By shaking the emulsion with successive volumes of chloroform, the phenylamine is gradually extracted into the organic layer. Since chloroform is immiscible with water, the organic layer is readily separated from the aqueous layer using a separating funnel.

4 Chemical Change

4.1 ENTHALPY CHANGES

When a reaction takes place, the chemical identity of the particles in the system changes: bonds are broken and new ones form in their place. Energy must be supplied in order to break the bonds, but it is given out when the new bonds form. In most reactions, these two energy factors are not equal, and so the whole system either gives out energy or takes in energy from the surroundings. Some reacting systems also do work on their surroundings.

> The total energy change between a reacting system and its surroundings is called the enthalpy change for the reactions, ΔH. The quantity ΔH equals the heat exchanged when the reaction takes place at constant pressure.

ΔH is given in units of kilojoules per mole (kJ mol^{-1}) and its sign indicates the direction of the heat exchange. A positive value of ΔH means that the system gains energy by taking it from the surroundings (an endothermic reaction), whereas a negative value of ΔH means that the system loses energy to the surroundings (an exothermic reaction).

Calorimetry

The experimental methods for measuring enthalpy changes are called calorimetry. There are two common types of calorimeter used.

1. The simple calorimeter for reactions in solution.
2. The bomb calorimeter for reactions involving gases.

Simple calorimeter

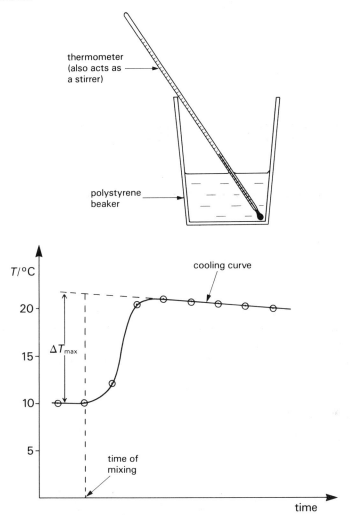

The simple calorimeter can be used as follows to determine ΔH for the displacement reaction between zinc dust and copper sulphate solution.

$$Zn(s) + Cu^{2+}(aq) \longrightarrow Zn^{2+}(aq) + Cu(s)$$

Firstly, the energy required to raise the temperature of the calorimeter system by 1°C needs to be measured (the 'calorimeter constant'). 20 grams of hot water at 40°C are poured into 20 grams of cold water at 10°C already in the calorimeter. The mixture is stirred and values of temperatures are recorded every half minute. A cooling-curve is plotted as shown above, and the curve is used to work out the temperature that the mixture would have reached in the absence of cooling losses. Extrapolation of the curve to the time of mixing gives this temperature as 22°C. In other words, the temperature of the hot water falls 18°C, while that of the cold water rises 12°C. When the temperature of a substance of mass m, and specific heat capacity s, changes by ΔT, the heat exchanged is given by the relationship:

$q = ms\Delta T$ ($s = 4.2 \text{ J g}^{-1}\text{ K}^{-1}$ for water)

∴ heat lost by hot water = $(20 \times 4.2 \times 18) = 1512$ J

∴ heat gained by cold water = $(20 \times 4.2 \times 12) = 1008$ J

The difference between these values is equal to the heat gained by the calorimeter system.

∴ heat gained by the calorimeter system = 504 J

But this amount of energy raised the temperature of the calorimeter system by 12°C, hence the energy required per degree rise = $(504/12) = 42$ J.

The value of the calorimeter constant (42 J °C^{-1}) is now used in the determination of ΔH for the displacement reaction.

1. 50 cm³ of the copper sulphate solution (0.1 mol dm^{-3}) are poured into the calorimeter and stirred with the thermometer.
2. Zinc powder is added by the spatula-full while the temperature is recorded every fifteen seconds.
3. When the temperature stops rising, no more zinc is added, but the temperature measurements are continued.
4. Cooling corrections are carried out as before and the corrected temperature rise is determined. In this case it is about 4.3°C.
5. The heat evolved by the reaction is, therefore, approximately equal to $(50 \times 4.2 \times 4.3) + (42 \times 4.3) = 1083$ J. The two terms are the heat energy gained by the solution and by the calorimeter respectively.
6. In 50 cm³ of 0.1 mol dm^{-3} CuSO$_4$, there is $(50/1000 \times 0.1) = 5 \times 10^{-3}$ mole. Since ΔH is the energy exchange per mole, this is given by $1083/(5 \times 10^{-3}) = 217\,000$ J. i.e. $\Delta H = -217$ kJ mol^{-1}.

Bomb calorimeter. The bomb calorimeter is used principally to measure enthalpies of combustion.

the bomb

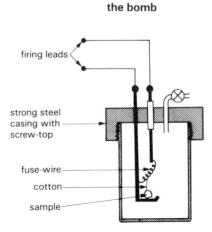

the calorimeter

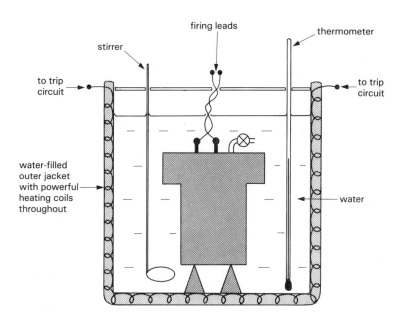

1. A known mass of the substance to be burnt is rolled into a pellet (if solid) and attached to a small piece of cotton.
2. The cotton is connected to a fuse wire that shorts the two terminals shown above.
3. The bomb is screwed up and filled to about 2 MPa with pure oxygen.
4. It is put into the calorimeter and the fuse is blown by a short pulse of current. This ignites the sample.
5. As the temperature rises in the calorimeter, the temperature of its outer jacket rises in step, because current flows through the heating-coils. The current for the coils is supplied from a trip circuit that is only triggered when the temperature of the calorimeter and the temperature of the outer jacket are different.
6. No cooling losses occur because the temperature of the surroundings remains the same as that of the reacting system. ΔT is measured directly.

The bomb needs to be calibrated so that a measured ΔT value can be converted into a ΔH value. This is usually done by burning a substance of standard ΔH: benzoic acid has a heat of combustion of -3227 kJ mol^{-1}. Some typical results are worked below.

$$12.0 \text{ g glucose } (M_r = 180) \longrightarrow \Delta T = 3.61°C$$
$$12.2 \text{ g of benzoic acid } (M_r = 122) \longrightarrow \Delta T = 6.20°C$$
$$\therefore 1 \text{ mole of glucose} \longrightarrow \Delta T = (3.61 \times 180/12.0) = 54.15°C$$
$$\text{and 1 mole of benzoic acid} \longrightarrow \Delta T = (6.20 \times 122/12.2) = 62.00°C$$

Since the enthalpy of combustion of the acid is -3227 kJ mol^{-1}, that of glucose $= (-3227 \times 54.15/62.00) = -2818$ kJ mol^{-1}.

Standard enthalpy changes and enthalpy diagrams

It is impossible to determine the amount of enthalpy in an isolated system; it is only possible to measure a *change* in the enthalpy of the system. To try and obtain an isolated measurement of an enthalpy is like trying to obtain an isolated measurement of a height. The height of a mountain only has meaning when it is referred to a standard height: that of sea-level. In the same way, a standard enthalpy must be chosen so that the amount of enthalpy in all other systems can be compared with the standard.

The chosen standards are:

1. one mole of any element in its natural state,
2. at one atmosphere pressure (0.1 MPa),
3. at 298 K.

The symbol for standard enthalpy is $H^{\ominus}$ and the standard enthalpy of an element at 1 atm and 298 K is, therefore, zero. This conclusion is reached by comparing the enthalpy of the element with that of the standard: the difference is zero, and hence $H^{\ominus} = 0$ kJ mol^{-1} for all elements at 1 atm and 298 K.

If a compound forms from its elements, an enthalpy change occurs because bonds are broken and made. The heat exchanged per mole of compound formed is called the standard enthalpy of formation $(\Delta H_f^{\ominus})$ of the compound. This can be shown on an enthalpy diagram. For example, for octane C_8H_{18}:

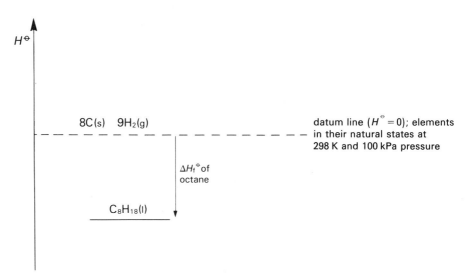

The standard enthalpy of formation of a compound $(\Delta H_f^{\ominus})$ is the enthalpy change when one mole of the compound is formed from its elements at 298 K and 0.1 MPa pressure.

Combustion processes are also easily represented on an enthalpy diagram. On the diagram below, the combusion of octane is given. The products are carbon dioxide and water which are also shown below as the combustion products of carbon and hydrogen.

The standard enthalpy of combusion of a substance ($\Delta H_c^\ominus$) is the enthalpy change when one mole of a substance is burnt in excess oxygen under standard conditions of temperature and pressure.

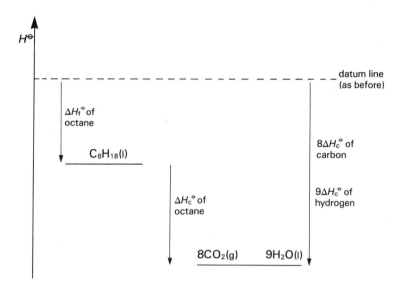

Hess's law. Hess applied the principle of the conservation of energy in the following way:

> Hess's law states that the enthalpy change for a process depends only on the initial and final states and not on the reaction pathway followed (providing the temperature and pressure remain constant).

For example, if it is possible that a reaction route in which A → B has $\Delta H = -80$ kJ while a second, different route has $\Delta H = -100$ kJ, then 20 kJ of energy could be destroyed by the following sequence. Hess's Law forbids this possibility.

$$A \xrightarrow[-80 \text{ kJ}]{\text{route one}} B \xrightarrow[+100 \text{ kJ}]{\text{route two}} A$$

Hess's law is used to calculate enthalpy changes that are not easy to measure directly. For example, it is impossible to determine directly $\Delta H_f^\ominus$ for octane: carbon and hydrogen do not combine to form octane. By applying Hess's law to the enthalpy data shown on the previous diagram,

$(\Delta H_f^\ominus + \Delta H_c^\ominus)$ of octane = $(8\Delta H_c^\ominus$ of carbon$) + (8\Delta H_c^\ominus$ of hydrogen$)$

The combustion data are easily measured by bomb calorimetry and so $\Delta H_f^\ominus$ for octane is found by subtraction.

Heats of reactions, $\Delta H_r^\ominus$. Any heat of reaction can be calculated by applying Hess's law to the standard enthalpies of formation of the reactants and products. For example, the enthalpy change per mole of calcium oxide for the slaking of lime,

$$CaO(s) + H_2O(l) \longrightarrow Ca(OH)_2(s)$$

is calculated as follows.

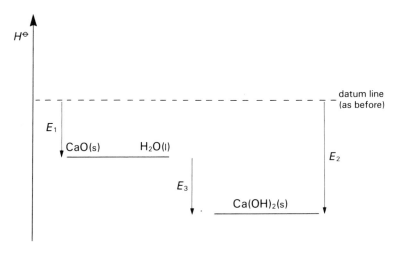

$E_1 = (\Delta H_f^\ominus \text{ of CaO}) + (\Delta H_f^\ominus \text{ of H}_2\text{O}) = -635 - 286 = -921 \text{ kJ}$

$E_2 = \Delta H_f^\ominus \text{ of Ca(OH)}_2 = -987 \text{ kJ}$

$E_3 = \Delta H_r^\ominus$ per mole of CaO

By Hess's law, $E_3 = E_2 - E_1 = -66 \text{ kJ}$

$\Delta H_r^\ominus = -66 \text{ kJ}$ per mole of CaO

4.2 BORN–HABER CYCLES

A Born–Haber cycle is an enthalpy diagram that relates the energy factors involved in the formation of a compound. These factors are derived from the bonding model adopted for the compound in question. For example, the covalent model is concerned with bond dissociation energies, while the ionic model is concerned with ionization energies, electron affinities and lattice energies (see pages, 7, 12 and 14).

The formation of an ionic compound

This is considered in terms of the relevant ionization energies, electron affinities and lattice energy. For example, the formation of sodium chloride is broken down into the following theoretical steps:

1 atomization of sodium and chlorine,
2 ionization of the atoms produced,
3 formation of an ionic lattice.

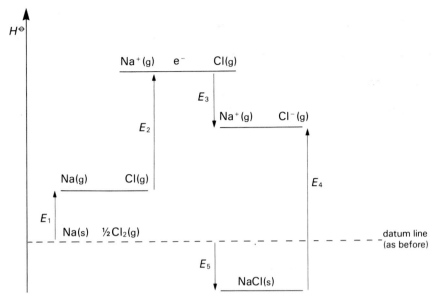

$E_1 = (\Delta H^{\ominus}_{at} \text{ of sodium}) + (\tfrac{1}{2}D^{\ominus} \text{ of Cl—Cl}) = 109 + 121 = 230 \text{ kJ}$

$E_2 =$ the first ionization energy of sodium $= 494$ kJ

$E_3 =$ the first electron affinity of chlorine $= -364$ kJ

$E_4 =$ the lattice energy of sodium chloride $= 771$ kJ

$E_5 = \Delta H^{\ominus}_f$ of sodium chloride

By Hess's law, $E_5 = E_1 + E_2 + E_3 - E_4 = -411$ kJ

∴ $\Delta H^{\ominus}_f$ of sodium chloride $= -411$ kJ mol^{-1}

The formation of a covalent compound

This is considered in terms of the bond dissociation energies needed to break the molecules into atoms. For example, the formation of sulphur dioxide assumes the molecular structures as shown below.

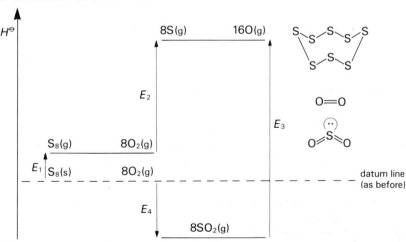

E_1 = the heat of sublimation of one mole of S_8 molecules = 102 kJ
$E_2 = (8D^\ominus \text{ of } S\text{—}S) + (8D^\ominus \text{ of } O=O) = 2112 + 3968 = 6080$ kJ
$E_3 = 16D^\ominus$ of $S=O$ = 8544 kJ
$E_4 = 8\Delta H_f^\ominus$ of SO_2

By Hess's law, $E_4 = E_1 + E_2 - E_3 = -2362$ kJ

∴ $\Delta H_f^\ominus$ of sulphur dioxide = -295 kJ mol^{-1}

The agreement between an experimentally determined and a theoretical value of $\Delta H_f^\ominus$ is not always good. Where there is disagreement, the assumptions made about the bonding of the compound must be questioned. For example, $\Delta H_f^\ominus$ for benzene is found experimentally from combustion data by the method described on page 61. But when the ring structure shown below is assumed, the theoretical value of $\Delta H_f^\ominus$ is 160 kJ mol^{-1} greater than the real value.

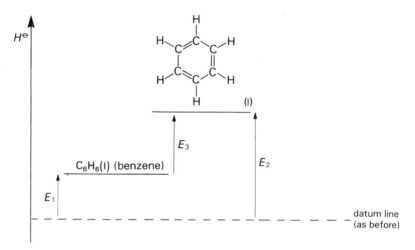

$E_1 = \Delta H_f^\ominus$ of benzene obtained from combustion data

$E_2 = \Delta H_f^\ominus$ of benzene predicted by bond dissociation energies

E_3 = the energy required to convert the molecular structure of benzene into the formula shown above (~160 kJ mol^{-1})

The true structure of a benzene molecule is more stable than that predicted by the above formula. The extra bond energy is called resonance energy or delocalization energy, and it is caused by the multiple π-overlap of the $2p_y$-orbitals of the carbon atoms (see page 194).

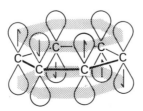

each carbon atom has a single-bonded hydrogen atom (not shown)

4 Chemical Change

Resonance. The benzene structure can, in fact, be represented by more than one structural formula. The two main ones are shown below.

once again, the hydrogen atoms have been left out

These two different structures are called the *canonical forms* of benzene. In describing the way in which any unsaturated particle is bonded, the use of canonical forms provides an alternative method of showing the effects of resonance or delocalization. Each canonical form illustrates an extreme case of the bonding present; the larger the number of canonical forms that can be drawn for a particular structure, the greater is the resonance effect, and the more stable the structure. For example, three canonical forms of the nitrate(v) ion can be drawn:

The resonance effect can also be shown by a π-overlap diagram:

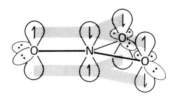

Although each canonical form is different, resonance theory does *not* imply that the bonding is alternating between one form and another. Instead, the structure is seen as a *resonant hybrid* of the different canonical forms, because it contains some of the character of each. For example, the resonant hybrid of benzene is often shown as follows.

Heats of solution, $\Delta H^{\ominus}_{sol}$

The most common laboratory solutions contain ions surrounded by water molecules. Although the principles outlined below concern the solubility of an ionic solid in water, they can be applied to any solute and solvent. The energy factors to be considered are:

1 the disruption of the lattice,
2 the surrounding of each solute particle by a sheath of solvent particles.

The second of the two factors is known as solvation or, when the solvent is water, hydration. The symbol for standard enthalpy of hydration is $\Delta H^\ominus_{hyd}$.

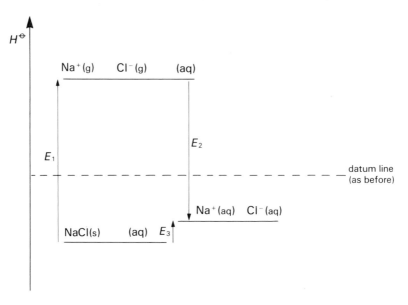

E_1 = the lattice energy of sodium chloride = 771 kJ

$E_2 = (\Delta H^\ominus_{hyd} \text{ of } Na^+) + (\Delta H^\ominus_{hyd} \text{ of } Cl^-) = -770$ kJ

E_3 = the heat of solution of sodium chloride, $\Delta H^\ominus_{sol}$

By Hess's law, $E_3 = E_1 + E_2$

∴ $\Delta H^\ominus_{sol}$ of sodium chloride = 1 kJ mol^{-1}

5 Reaction Rates

5.1 MEASURING RATES

Factors that affect rate

The rate of a reaction is a measure of the amount of reactants being converted to products in a unit time. It is expressed as the concentration of reactants consumed per unit time or products generated per unit time. For example, in the reaction $A \rightarrow B$, the rate is either $-d[A]/dt$ or $d[B]/dt$, where d/dt is the mathematical symbol meaning 'the rate of change of'. In most cases, the units for rate are $mol\ dm^{-3}$ of reactant or product per second: $mol\ dm^{-3}\ s^{-1}$.

The rate of a reaction can be shown to depend on:

1. the concentration of reactants present,
2. the temperature,
3. the pressure (for gaseous reactions),
4. the presence of a catalyst,
5. the presence of electromagnetic radiation (for photochemical reactions).

The 'discontinuous' method

The dependence of rate on reactant concentration can be established by using a 'discontinuous' method. For example, in the reaction $A + B + C \rightarrow$ products,

1. different concentrations of reactants are measured into several different reaction flasks at a constant temperature in a water bath;
2. the concentrations of two of the reactants are kept constant from flask to flask while that of the other one is varied from flask to flask;
3. a method for determining the initial rate of the reaction is chosen: usually a small sample is withdrawn after a certain interval of time and analysed for a particular reactant or product;
4. the initial rates in each flask are then compared with the initial reactant concentrations present and their relationship is, therefore, deduced;

5 the procedure is repeated so that each reactant concentration is varied in turn. The dependence of the rate on [A], [B] and [C] can then be expressed as a 'rate law' in the following form (where a, b and c usually have a value of either 0, 1 or 2).

Rate $\propto [A]^a[B]^b[C]^c$

Rate $= k[A]^a[B]^b[C]^c$ where k is a proportionality constant.

> The constant of proportionality in the rate law for a particular reaction is called the rate constant or reaction velocity constant, k.

The power to which each reactant concentration is raised in the rate law (a, b and c in the equation), is known as the individual order of the reaction with respect to the reactant concerned. The overall order of the reaction is the sum of the individual orders.

> The order of a reaction is the sum of the powers to which the reactant concentrations are raised in the rate law for the reaction.

For example, the initial rate of the reaction between propanone and iodine in the presence of acid can be found by sampling the mixture after a short interval and then analysing for iodine using sodium thiosulphate (page 143).

$$CH_3COCH_3 + I_2 \xrightarrow{H_3O^+} CH_3COCH_2I + HI$$

When the initial concentrations of iodine and acid are kept constant in the different flasks, the rate of consumption of iodine is found to triple if the propanone concentration is tripled, i.e. rate $\propto$ [propanone]^1. Had the rate increased by a factor of nine when the concentration of propanone tripled, the rate would have been proportional to [propanone]^2.

When the initial concentrations of propanone and acid are kept constant, however, it is found that the rate of iodine consumption is unaffected by increasing the initial concentration of iodine i.e. rate $\propto [I_2]^0$, or rate = a constant.

The rate law indicates that the overall order is two. Although the reaction is zero order with respect to iodine, it is first order with respect to both propanone and the acid catalyst.

$$\text{Rate} = k[CH_3COCH_3]^1[I_2]^0[H_3O^+]^1 \qquad \text{order} = (1 + 0 + 1) = 2$$

The mechanism of the reaction is discussed on page 214.

The 'continuous' method

For certain types of reaction, the overall order can be found by monitoring the progress of a single reaction mixture to its completion. Reactions of this sort include decomposition processes and reactions whose reactant stoichiometry is in a 1:1 ratio. These are illustrated by the following equations.

$A \longrightarrow$ products   decomposition

$A + B \longrightarrow$ products $A:B = 1:1$

If c stands for reactant concentration, the rate of the reaction is $-dc/dt$. Providing [A] = [B] in reaction mixtures of the above type, the first and second order rate laws then reduce to the form shown in the table below. The integrated forms are also given; c_0 is any chosen value of c, and c_t is the new value of c after a time interval of t.

order	rate law	integrated rate law
first	$-\dfrac{dc}{dt} = kc$	$\log_e \dfrac{c_0}{c_t} = kt$
second	$-\dfrac{dc}{dt} = kc^2$	$\dfrac{1}{c_t} - \dfrac{1}{c_0} = kt$

By sampling a reaction mixture at different intervals of time and analysing each sample for reactant concentration, a set of values of c_t at different values of t is obtained. These values can be used to show whether the reaction is first or second order.

procedure	observation	conclusion
plot $\log_e (c_t)$ against t	the graph is a straight line of negative slope	the data fits the first order rate law
plot $1/c_t$ against t	the graph is a straight line of positive slope	the data fits the second order rate law

The gradient of the slope in either case gives the value of the rate constant. For example, the order and rate constant of the reaction of an ester with alkali are found as follows.

1. A known mass of ester is added to an equimolar amount of alkali of known concentration in a water-bath at 25°C.
2. Small samples are pipetted out of the mixture on a number of separate occasions, noting the reaction time t on each occasion.
3. Each sample is 'frozen' by pouring it into a large volume of cold water. This has the effect of slowing the rate down almost to zero.
4. Each frozen sample is titrated against standard acid and the concentration of alkali [OH$^-$], is found in each case.
5. Although a plot of $\log_e$ [OH$^-$] against t does *not* give a straight line, a plot of $1/$[OH$^-$] against t does. This suggests that the reaction is second order and the gradient of the slope gives the value of the rate constant.

Half-life method

The half-life, $t_{1/2}$, of a reaction is the time taken for the initial reactant concentration c_0, to fall to half its value, $c_0/2$.

If the above conditions, $t = t_{1/2}$ when $c_t = c_0/2$, are applied in the two integrated rate law expressions, the following results are obtained.

first order	second order
$\log_e \left(\dfrac{c_0}{c_t} \right) = kt$	$\dfrac{1}{c_t} - \dfrac{1}{c_0} = kt$
$t = t_{1/2}$ when $c_t = c_0/2$	$t = t_{1/2}$ when $c_t = c_0/2$
$\therefore \log_e \left(\dfrac{c_0}{c_0/2} \right) = kt_{1/2}$	$\therefore \dfrac{1}{c_0/2} - \dfrac{1}{c_0} = kt_{1/2}$
$\therefore \log_e 2 = kt_{1/2}$	$\therefore \dfrac{1}{c_0} = kt_{1/2}$
or $t_{1/2} = \dfrac{\log_e 2}{k}$, a constant	or $c_0 t_{1/2} = k$, a constant

It is, therefore, possible to find the order and rate constant of a reaction by inspecting a plot of reactant concentration against time. The half-life is measured from several different initial reactant concentrations. If the values are constant, the reaction is likely to be first order and the rate constant is $(\log_e 2)/t_{1/2}$. If the values increase steadily but the product $c_0 t_{1/2}$ remains constant, the reaction is likely to be second order and the

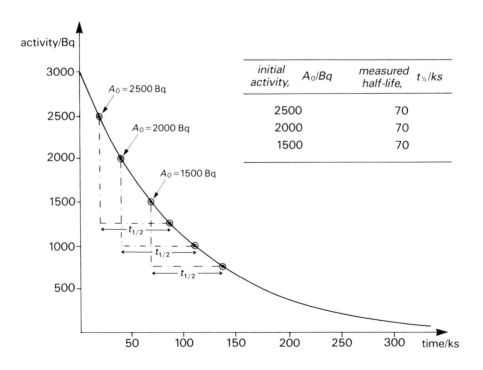

initial activity, A_0/Bq	measured half-life, $t_{1/2}$/ks
2500	70
2000	70
1500	70

rate constant is $1/(c_0 t_{1/2})$. For example, radioactive decay (page 3) can be shown to obey first order kinetics by the following method.

1. The level of radioactivity of a sample is measured at various times using a Geiger counter.
2. The amount of radioactive matter is assumed to be proportional to the counts per second (bequerels) recorded at the different decay times.
3. A plot of count rate against time shows that the half-life is a constant. In the curve on the facing page, $t_{1/2} = 7.0 \times 10^4$ s.
4. The decay is, therefore, first order and the rate constant is given by $(\log_e 2)/t_{1/2}$. For a decay process, the rate constant is called the decay constant, λ.

i.e. $\lambda t_{1/2} = \log_e 2 = 0.693$

$\therefore \lambda = 9.9 \times 10^{-6}\, s^{-1}$

5.2 COLLISION THEORY

The collision theory was developed to explain the dependence of rate on reactant concentration and temperature. Reaction mechanisms (page 178) are also an application of the theory.

Assumptions

1. A reaction occurs as a result of a collision between reactant particles.
2. Not every collision leads to a reaction. There is a minimum energy associated with an effective collision.
3. When this minimum energy is summed for a mole of collisions, it is called the activation energy, E_A.

> The activation energy, E_A, is the minimum energy required in a collision so that the collision results in a reaction: E_A is expressed per mole of collisions.

4. Even when the energy requirements are satisfied, a collision may still be ineffective if the particles are not correctly orientated with respect to each other. Every reaction has a steric factor, p, which is the fraction of collisions that satisfy the orientation requirements $(0 < p < 1)$.

Using these assumptions, the rate of a reaction can be expressed as shown below.

$$\text{rate} = \frac{\text{the number of effective collisions happening}}{\text{per second in a unit volume of the system.}}$$

By defining a 'collision number', Z, as the average number of collisions that each individual reactant particle undergoes per second with other reactant particles present in the system, the above expression takes the following form.

$$\text{rate} = e^{-E_A/RT} \times p \times \text{reactant concentrations} \times Z$$

$e^{-E_A/RT}$ is the fraction of collisions whose energy per mole exceeds the activation energy, E_A, at the particular temperature, T. p is the fraction of the collisions with the correct orientation, and the reactant concentration factors take into account the whole population of particles present per unit volume.

At a constant temperature, the value of $(pZ e^{-E_A/RT})$ is constant and hence the expression reduces to the rate law introduced earlier:

rate = k × reactant concentrations at constant T

where $k = pZ e^{-E_A/RT}$

The constants pZ are often replaced by a single symbol, A, which is called the pre-exponential factor. In this form, the relationship equating the rate constant to temperature and activation energy is named after Arrhenius who derived it from experimental data.

Arrhenius equation: $k = A e^{-E_A/RT}$ or $\log_e k = \log_e A - E_A/RT$

Applications

The collision theory explains the dependence of rate on reactant concentration and temperature as follows.

Reactant concentration. An increase in the concentration of a reactant leads to an increase in the frequency of collisions. In turn, this produces an increase in the frequency of effective collisions and, therefore, an increase in rate.

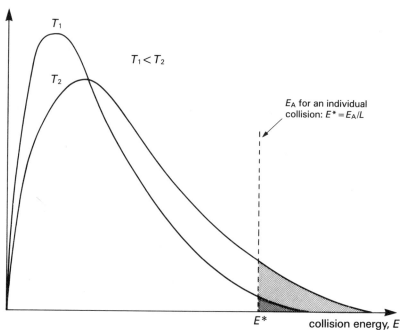

Temperature. An increase in temperature leads to an increase in the average energy of the collisions between the reactant particles. At a higher temperature, a higher proportion of the collisions satisfy the energy requirements because this proportion, $e^{-E_A/RT}$ increases with temperature, T. In consequence, the rate increases rapidly on raising the temperature of a reaction.

The effect is illustrated on the previous page by looking at the distribution of collision energies in a sample at two different temperatures. The variation is a form of the Maxwell–Boltzmann function shown on page 31 for the distribution of molecular speeds. The difference in the shape of the two sets of curves reflects that energy is proportional to (speed)^2.

The total area under each curve is a measure of the total number of collisions taking place per second. The shaded area represents the fraction of collisions at the lower temperature that satisfy the energy requirements. The hatched area represents those that are effective at the higher temperature. There is a marked increase in the fraction as the temperature increases.

5.3 MULTI-STEP REACTIONS

Rate-determining step

Many reactions take place in a series of simple steps. For example, the iodination of propanone proceeds via the *enol*-form of propanone, but over 99% of the propanone is present as the *keto*-form (see page 211). The reaction is a two-step process:

The rate at which *enol*-propanone forms is very much slower than its subsequent iodination, so that this second step is almost instantaneous by comparison with the rate of the first step. In other words, the overall rate depends only on the rate of the first step. For this reason, the slow conversion of *keto*-propanone to *enol*-propanone is called the *rate-determining step*.

The concentration of iodine present cannot affect the overall rate because iodine does not take part in the rate-determining step. On the other hand, however, an increase in the concentration of propanone leads to a subsequent increase in the rate of the rate-determining step. Hence, the rate law has the following form.

rate $\propto$ [propanone]^1[iodine]^0

The rate-determining step of any multi-step process (except a chain reaction) is always the slowest in the series. Sometimes this is the first step, as in the above example; sometimes it is the final one (see page 78).

There are two important classes of multi-step process: catalysed reactions and chain reactions.

Catalysed reactions

A catalyst is a substance that speeds up the rate of a chemical reaction but is itself chemically unaltered at the end of the reaction.

The action of a catalyst is best understood by considering its effect on the activation energy of the reaction. In bulk terms, the activation energy is an energy barrier that the reactants must overcome in order to react. It is the energy difference between the reactants and the activated complex.

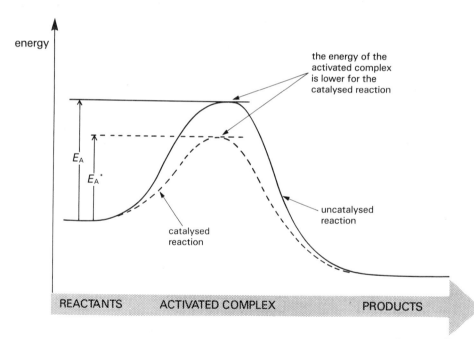

An activated complex cannot be isolated from its reaction mixture. It contains reactant particles that have undergone partial bond rupture and formation as a result of a collision. For example, in the alkaline hydrolysis of bromoethane (see page 179):

A catalyst has one of the following two functions.

1. It makes it easier for the activated complex to form and, hence, lowers the activation energy.
2. It provides an alternative reaction pathway via a different activated complex of lower energy.

The rate constant increases with decreasing activation energy (according to the Arrhenius equation, $k = A\,e^{-E_A/RT}$), so the lower the activation energy, the faster the rate. In both the above catalytic functions, the catalyst lowers the activation energy. There are three types of catalyst: homogeneous catalysts, heterogeneous catalysts and enzymes.

Homogeneous catalysts. These function in the same phase as the reactants and a different activated complex of lower energy is usually formed. The catalysed reaction is a multi-step process in which the catalyst is used up in an initial step and regenerated in a later one. For example, the iodination of propanone described earlier is catalysed by acid, and the full rate law is shown below.

$$\text{rate} \propto [\text{propanone}]^1 [\text{iodine}]^0 [\text{acid}]^1$$

Step 1 is slow and rate-determining (see page 75): the acid catalyst speeds up the rate of the *keto-enol* conversion as follows.

Step 2 is almost instantaneous compared with the rate of step 1.

Heterogeneous catalysts. These function in a different phase from the reactants and usually assist the formation of the same activated complex that forms in the uncatalysed reaction. The whole process takes place at the catalyst phase boundary and this has two effects.

1. The reactant particles become partially bonded to the catalyst and this weakens the bonds that are broken during the reaction.
2. The reactant particles are brought closer together at the catalytic surface than they are in the reaction mixture.

Both these effects contribute to a lowering of the activation energy. For example, the decomposition of ammonia on tungsten is a zero-order reaction: rate $\propto$ [ammonia]^0. The tungsten surface acts as a heterogeneous catalyst in a multi-step reaction whose rate-determining step does not depend on the concentration of ammonia present.

Step 1, the adsorption of ammonia, is very fast.

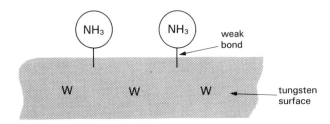

$NH_3(g) \longrightarrow NH_3(on\ W)$

Step 2, the reaction at the surface, is also comparatively fast.

$2NH_3(on\ W) \longrightarrow N_2(on\ W) + 3H_2(on\ W)$

Step 3, the desorption of products, is very slow compared with the rate of steps 1 and 2; it is rate-determining, for example,

$H_2(on\ W) \longrightarrow H_2(g)$

The rate-determining step does not depend on the concentration of ammonia but only on the temperature and the affinity of the products for tungsten. Hence, the overall rate law is given by: rate $\propto$ $[NH_3]^0$.

Enzymes. These are macromolecular organic substances whose catalytic functions combine those of both homogeneous and heterogeneous catalysts. Most enzymes operate in the same phase as their reactants but possess a molecular chain that has the property of forming weak bonds to the reactant particles. There are 'active sites' at particular positions along the chain of the enzyme and these are called templates. For example, blood contains enzymes which catalyse the decomposition of hydrogen peroxide to water and oxygen. These catalysts prevent the accumulation of peroxide in the blood-stream as a by-product of respiration.

Chain reactions

A chain reaction is a multi-step process in which the rate-determining step is not the slowest one. Most chain reactions involve free radicals as 'chain carriers'; a free radical is an atom or a bonded group containing a single unpaired electron. For example, a methyl and a chlorine radical are shown below.

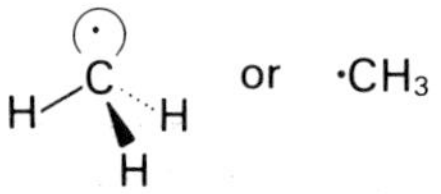

methyl radical 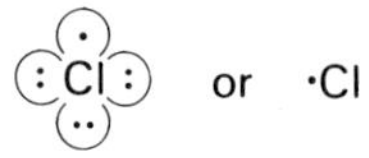   chlorine radical

Chain reactions have three distinguishable stages: initiation, propagation and termination. For example, in the photochemical reaction of chlorine and methane, no reaction occurs unless it is initiated by ultraviolet radiation.

Initiation. A chlorine molecule undergoes homolytic fission (see page 178) by absorbing energy from the ultraviolet radiation.

$$Cl-Cl \xrightarrow{h\nu} Cl\cdot \quad \cdot Cl$$

chlorine molecule → chlorine radicals

Initiation processes have a high activation energy and are slow.

Propagation. The collision of a free radical and a reactant molecule leads to the formation of a product molecule and a different radical. This new radical collides with another molecule and the process repeats itself. There are a huge number of these propagation steps for every initiation that occurs: a chain is set up. For example,

$$Cl\cdot + H-CH_3 \longrightarrow Cl-H \quad \cdot CH_3$$

$$Cl-Cl + \cdot CH_3 \longrightarrow Cl\cdot \quad Cl-CH_3$$

Propagation processes have a very low activation energy and are extremely rapid.

Termination. The propagation of the chain is terminated when two radicals collide and join together. This can only happen if the colliding radicals manage to lose energy by colliding with a third particle before they split apart again. Without this second collision the radicals have enough energy to bounce off each other.

$$Cl\cdot + \cdot CH_3 \xrightarrow[\text{body (e.g. the vessel wall)}]{\text{in collision with a third}} Cl-CH_3$$

Chain reactions are unspecific and a large variety of different products are often produced, for example:

$$CH_4 \xrightarrow[\text{sunlight}]{\text{chlorine}} CH_3Cl, CH_2Cl_2, CHCl_3, CCl_4 + HCl$$

6 Chemical Equilibrium

6.1 THE EQUILIBRIUM LAW

Reactions in solution, K_c

A reversible reaction reaches equilibrium when the rate of the forward reaction equals the rate of the back reaction. Under these conditions, there is no change in the concentration of either reactants or products. For example, in the equilibrium mixture represented below,

$$a\text{A} + b\text{B} \rightleftharpoons c\text{C} + d\text{D}$$
$$\text{reactants} \qquad \text{products}$$

the equilibrium concentrations [A], [B], [C] and [D] are related by the following expression, known as the equilibrium law.

$$K_c = \frac{[C]^c[D]^d}{[A]^a[B]^b} \qquad \text{where } K_c \text{ is a constant called the equilibrium constant.}$$

The relative amount of products and unconsumed reactants in an equilibrium mixture is summarized by the term position of equilibrium, shown in the table opposite.

Measuring K_c. The acid-catalysed hydrolysis of an ester (page 226) is a typical aqueous reaction that does not go to completion. The equilibrium constant is measured as follows.

$$\text{RCOOR} + \text{H}_2\text{O} \xrightleftharpoons{\text{H}_3\text{O}^+} \text{RCOOH} + \text{ROH}$$
$$\text{ester} \qquad \text{water} \qquad \text{carboxylic} \quad \text{alcohol}$$
$$\text{acid}$$

$$K_c = \frac{[\text{RCOOH}][\text{ROH}]}{[\text{RCOOR}][\text{H}_2\text{O}]}$$

value of K_c	reactants	equilibrium position	products
$K_c < \sim 10^{-2}$	there is a greater proportion of reactants at equilibrium: the position of equilibrium lies to the left		
$K_c > \sim 10^{2}$			there is a greater proportion of products at equilibrium: the position of equilibrium lies to the right

1. Make up several different mixtures of ester, water and alcohol using known amounts of each substance.
2. Measure out a small, fixed amount of concentrated sulphuric acid into each flask as catalyst.
3. Cork the flasks and leave them in a water-bath for a week to equilibrate at constant temperature.
4. Analyse the equilibrium mixtures for the concentration of acid present: by titration against standard alkali the number of moles of acid can be found in each case.
5. Subtract the number of moles of acid used as a catalyst and hence, find the concentration of the carboxylic acid at equilibrium in each case.
6. The other equilibrium concentrations are calculated from the known starting amounts: the stoichiometry of the reaction shows that for every mole of acid formed, one of alcohol is formed and one each of ester and water are used up.
7. The equilibrium concentrations are put into the above expression for K_c. An average value is taken.

Reactions in the gas phase, K_p

The concentration of a gaseous reactant in an equilibrium mixture is directly proportional to the pressure it exerts. This relationship follows directly from the ideal gas law, $pV = nRT$ because the concentration equals n/V, the number of moles per unit volume.

$pV = nRT$

$n/V = (1/RT)p$

If there is more than one gas present, then the pressure exerted by component A is given by its partial pressure, p_A (see page 31). The above expression now becomes:

$[A] = (1/RT)p_A$

A different equilibrium constant, K_p is defined if the equilibrium partial pressures are applied in the equilibrium law. For example, both K_c and K_p can be used to describe the equilibrium shown below.

$$N_2(g) + 3H_2(g) \rightleftharpoons 2NH_3(g)$$

$$K_c = \frac{[NH_3]^2}{[N_2][H_2]^3} \qquad K_p = \frac{p_{NH_3}^2}{p_{N_2} \times p_{H_2}^3}$$

In this case, since each of the concentrations can be written in the form $(1/RT)p$,

$$K_c = \frac{(1/RT)^2 p_{NH_3}^2}{(1/RT)p_{N_2} \times (1/RT)^3 p_{H_2}^3} = \frac{K_p}{(1/RT)^2}$$

$$\therefore K_c = (RT)^2 K_p$$

Factors that affect the position of equilibrium

The position of an equilibrium is affected by a change in:

1 the concentration of a reactant or product,
2 the pressure (for gas),
3 the temperature.

A catalyst has no effect on the position of an equilibrium because it speeds up the rate of both the forward and the backward reaction by the same factor. So, although a catalyst enables a reaction mixture to reach equilibrium more quickly, it cannot be used as a means of shifting the equilibrium position.

Concentration. If the concentration of a reactant or a product in an equilibrium mixture is suddenly changed, the position of equilibrium shifts until equilibrium is re-established. For example, during the synthesis of ammonia, a sudden increase in the concentration of hydrogen in an equilibrium mixture unbalances the equilibrium.

$$K_c = \frac{[NH_3]^2}{[N_2][H_2]^3} \qquad N_2(g) + 3H_2(g) \rightleftharpoons 2NH_3(g)$$

sudden increase

As $[H_2]$ increases, the values of $[N_2]$ and $[NH_3]$ must change in order that the value of K_c should remain constant. More ammonia is, therefore, produced at the expense of nitrogen so that $[NH_3]$ increases and $[N_2]$ decreases. In other words, the equilibrium position shifts to the right when extra reactant is added to an equilibrium mixture.

By a similar process of reasoning the addition of extra product to an equilibrium mixture can be shown to cause a shift to the left. Similarly, the sudden removal of a reactant or product from an equilibrium mixture has the opposite effect to that of its addition.

Pressure. For gaseous reactions, the equilibrium law is best expressed in partial pressures. For example, in an equilibrium mixture containing the following mole fractions of ammonia, nitrogen and hydrogen, x_{NH_3}, x_{N_2} and x_{H_2}, the partial pressures are:

$$p_{NH_3} = x_{NH_3} \times p_T$$

$$p_{N_2} = x_{N_2} \times p_T \qquad \text{where } p_T = p_{NH_3} + p_{N_2} + p_{H_2} \qquad \text{(Dalton's law)}$$

$$p_{H_2} = x_{H_2} \times p_T$$

$$\therefore K_p = \frac{x_{NH_3}^2}{x_{N_2} \times x_{H_2}^3} \times \frac{1}{p_T^2}$$

A sudden increase in pressure unbalances the equilibrium because $1/p_T^2$ gets *much* smaller in the above expression. In order that the value of K_p should remain constant, the mole fractions present in the system must change: x_{NH_3} increases at the expense of x_{N_2} and x_{H_2} until the decrease in the value of $1/p_T^2$ is counterbalanced. In other words, the increase in pressure causes the position of equilibrium to shift to the right. Similarly, a sudden decrease in pressure would shift the equilibrium position to the left. Low pressures favour the side with the larger number of moles; high pressures favour the side with the smaller number of moles.

Temperature. On raising the temperature of an equilibrium mixture, the rates of both forward and backward reactions increase. However, the factor by which each rate increases is very rarely the same for both (as *is* the case when a catalyst is added, page 76). An exothermic forward reaction must have an opposing backward reaction that is endothermic. For example,

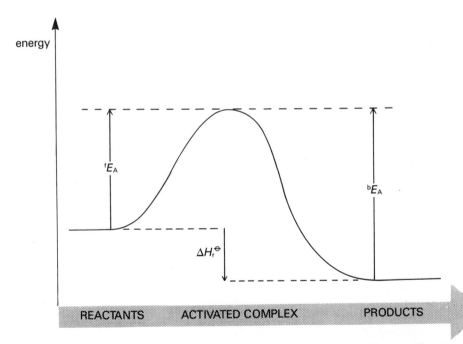

Since the rate constant of any reaction, k is related to its activation energy, E_A by the Arrhenius equation, $\log_e k = -E_A/RT + \log_e A$, it therefore follows that the larger the value of E_A, the larger is the increase in $\log_e k$ when T is increased.*

In the above diagram $^bE_A > {}^fE_A$ and hence an increase in temperature leads to a greater increase in the rate of the backward reaction compared with that of the forward reaction. The equilibrium position, therefore, shifts to the left on increasing the temperature, and K_p *actually changes in value* and gets smaller. In other words, an increase

* $\dfrac{d(\log_e k)}{dT} = \dfrac{E_A}{RT^2}$ by differentiation with respect to T

in temperature favours the endothermic reaction, while a decrease favours the exothermic reaction.

The exact relationship between K_p, T and $\Delta H_r^\ominus$, the standard enthalpy change for the forward reaction, is given by the following equation.

$$\frac{d(\log_e K_p)}{dT} = \frac{\Delta H_r^\ominus}{RT^2} \quad \text{(the van 't Hoff equation)}$$

When $\Delta H_r^\ominus$ is positive (endothermic forward reaction), K_p increases as the temperature increases; when $\Delta H_r^\ominus$ is negative (exothermic forward reaction), K_p decreases as the temperature increases.

Le Chatelier's principle. Le Chatelier summarized the effects of a sudden change in the conditions of an equilibrium system. His principle is not an attempt to explain the effects, but is merely a guide for predicting the direction of any shift in equilibrium position.

> Le Chatelier's principle states that when an external constraint is applied to an equilibrium mixture the equilibrium position shifts in such a way that the effect of the constraint is minimized.

For the three constraints already discussed, the principle is applied as follows.

1. A sudden increase in reactant concentration: the equilibrium position shifts to the right in order to use up as much of the added reactant as possible.
2. A sudden increase in the pressure of a gaseous equilibrium: the equilibrium position shifts to the side with the smaller number of moles which, by Avogadro, has the smaller volume. This minimizes the effect of the increased pressure.
3. An increase in temperature: the equilibrium position shifts so that the maximum amount of heat can be absorbed i.e. the endothermic process is favoured.

6.2 PROTON TRANSFER

The Brønsted–Lowry theory

Acids and bases. According to Brønsted and Lowry, an acidic particle is a proton donor. It contains a hydrogen atom covalently bonded to a more electronegative atom. Under these conditions, the core of the hydrogen atom is a poorly-shielded proton (see page 20) and can be strongly attracted to the lone pair of another particle.

A basic particle is a proton acceptor. It has a lone pair of electrons able to form a covalent bond to the electron-deficient hydrogen atom of an acidic particle. For example,

acid   base

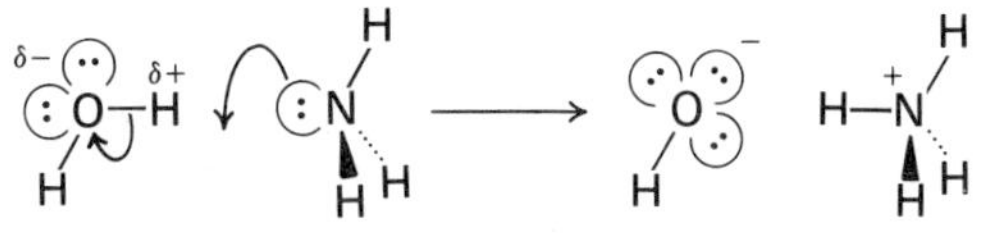

6 Chemical Equilibrium

In the first example, water acts as a base because it accepts a proton; in the second, it acts as an acid because it donates a proton. Water is capable of both acidity and basicity and this property is called amphoteric behaviour.

> Acidic particles are proton donors; basic particles are proton acceptors; amphoteric particles can act as both proton donors and acceptors.

Proton transfer. This is a reversible process. For the general acid, HA in water, two opposing proton transfer processes can occur.

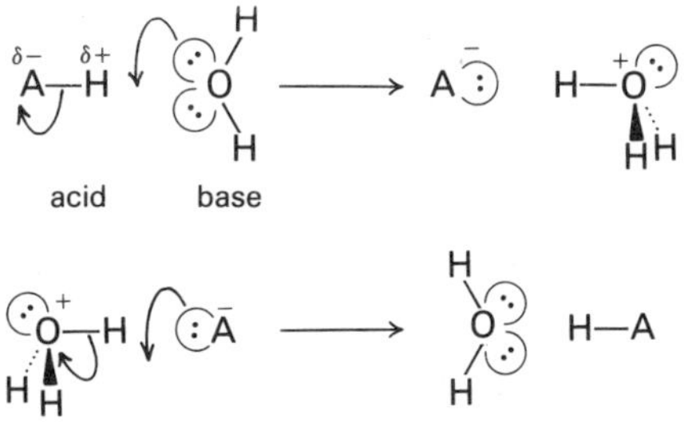

In the system, there are two acids and two bases.

$$HA(aq) + H_2O(l) \rightleftharpoons H_3O^+(aq) + A^-(aq)$$
acid 1   base 1   acid 2   base 2

Acid 1 is the protonated form of base 2, while base 1 is the deprotonated form of acid 2. The terms *conjugate acid* and *conjugate base* are used to describe this effect. Each acid has its conjugate base while each base has its conjugate acid.

conjugate acid	conjugate base
HA	A^-
H_3O^+	H_2O

If the equilibrium law is applied to the above process of proton transfer, the following expression is derived.

$$K_c = \frac{[H_3O^+][A^-]}{[HA][H_2O]} \quad \text{or} \quad K_c[H_2O] = \frac{[H_3O^+][A^-]}{[HA]}$$

In dilute solutions, the concentration of water is approximately constant ($1000/18$ mol dm^{-3}). Hence, the expression $K_c[H_2O]$ can be replaced by a single constant. It is given the symbol K_a and is called the dissociation constant of the acid.

$$K_a = \frac{[H_3O^+][A^-]}{[HA]}$$

The strength of an acid is determined by the position of the above equilibrium. When it lies well to the right ($K_a > 10^2$ mol dm^{-3}), the acid is strong; when it lies well to the left ($K_a < 10^{-2}$ mol dm^{-3}), the acid is weak.

An exactly analogous treatment can be applied to a weak base in water. The dissociation constant is given the symbol, K_b.

$$B(aq) + H_2O(l) \rightleftharpoons HB^+(aq) + OH^-(aq)$$

$$K_b = \frac{[HB^+][OH^-]}{[B]}$$

Water acts as both a weak acid and a weak base. When the above treatment is applied to the equilibrium below, the dissociation constant of water, K_w is derived. $K_w = 10^{-14}$ mol^2 dm^{-6} at 298 K and 0.1 MPa pressure.

$$H_2O(l) + H_2O(l) \rightleftharpoons H_3O^+(aq) + OH^-(aq)$$

$$K_w = [H_3O^+][OH^-]$$

Salts and their hydrolysis

A salt is formed by the reaction of an acid and a base. For example, a sodium salt is produced as follows.

$$HA + NaOH \longrightarrow NaA + H_2O$$

If both acid and base are strong, the salt is completely dissociated into aqueous ions in solution. The ions are fully hydrated (page 68) but are not hydrolysed: hydrolysis means 'reaction with water'. However, if either the acid or the base is weak, then the salt produced is hydrolysed in solution. Proton transfer takes place between the solvent water molecules and the dissociated ions from the salt.

Salts of weak acids. Typical weak acids include ethanoic acid and carbonic acid. Their salts are hydrolysed as follows.

$$NaA(s) \xrightarrow[\text{hydration}]{(aq)} Na^+(aq) + A^-(aq)$$

$$A^-(aq) + H_2O(l) \underset{\text{hydrolysis}}{\rightleftharpoons} \underset{\substack{\text{weak}\\\text{acid}}}{HA(aq)} + OH^-(aq)$$

The above reaction ensures that there are more hydroxide ions than hydroxonium ions in the solution: $[OH^-] > [H_3O^+]$. The solution is, therefore, alkaline with a pH greater than seven.

Salts of weak bases. Typical weak bases include ammonia and those metal hydroxides which contain small, highly-charged cations (e.g. Al^{3+}, Fe^{3+}). Their salts are hydrolysed as follows.

$$NH_4Cl(s) \xrightarrow[\text{hydration}]{\text{(aq)}} NH_4^+(aq) + Cl^-(aq)$$

$$NH_4^+(aq) + H_2O(l) \xrightleftharpoons[\text{hydrolysis}]{} NH_3(aq) + H_3O^+(aq)$$

$$AlCl_3(s) \xrightarrow[\text{hydration}]{\text{(aq)}} Al^{3+}(aq) + 3Cl^-(aq)$$

$$\left[(H_2O)_5Al \overset{H}{\underset{H}{\cdot\cdot O}}\right]^{3+} + H_2O \xrightleftharpoons[\text{hydrolysis}]{} \left[(H_2O)_5Al\!:\!\ddot{O}\!-\!H\right]^{2+} + H_3O^+$$

The production of hydroxonium ions means that $[H_3O^+] > [OH^-]$ in the solution which is, therefore, acidic (pH less than seven).

pH and acid strength

> The pH of a solution is defined as the negative of the logarithm (to base ten) of the hydroxonium ion concentration (in mol dm^{-3}) present in the solution.
>
> $$pH = -\log_{10}[H_3O^+] \quad \text{or} \quad [H_3O^+] = 10^{-pH}$$

Strong acids. The pH of a solution of a strong acid (for example, 0.1 mol dm^{-3} HCl) is calculated as follows.

1 Assume that the acid is fully dissociated.

   $HCl(aq) + H_2O(l) \longrightarrow H_3O^+(aq) + Cl^-(aq)$

2 Calculate $[H_3O^+]$ using the stoichiometry of the equation and the initial acid concentration.

   1 mole HCl $\longrightarrow$ 1 mole H_3O^+

   $\therefore$ 0.1 mol dm^{-3} HCl $\Longrightarrow$ $[H_3O^+]$ = 0.1 mol dm^{-3}

3 Convert the value of $[H_3O^+]$ to a power of ten and apply the definition of pH.

   $[H_3O^+] = 0.1 = 10^{-1}$ mol dm^{-3}

   $\therefore$ pH = 1

Weak acids. The pH of a solution of a weak acid (for example, 0.1 mol dm^{-3} ethanoic acid, HA) is calculated as follows.

1 Write down the equilibrium of HA in water and assume that x mol dm^{-3} of H_3O^+ are formed at equilibrium.

$$HA(aq) + H_2O(l) \rightleftharpoons H_3O^+(aq) + A^-(aq)$$

	HA(aq)	H_3O^+	A^-	
initial amount/mol	0.1	0	0	in 1 dm^3
equilibrium amount/mol	$(0.1 - x)$	x	x	in 1 dm^3

2 Write down the equilibrium concentrations in terms of x, and relate these to the value of K_a for the acid (K_a for ethanoic acid = $10^{-4.8}$ mol dm^{-3}).

$$[HA] = (0.1 - x) \text{ mol dm}^{-3} \quad [H_3O^+] = [A^-] = x \text{ mol dm}^{-3}$$

$$K_a = \frac{[H_3O^+][A^-]}{[HA]} \implies \frac{x^2}{0.1 - x} = 10^{-4.8}$$

3 Assume that the proportion of the weak acid that dissociates is so small that $[HA]_{initial} \approx [HA]_{equilibrium}$. Under this assumption, $0.1 \approx (0.1 - x)$.

$$\therefore \frac{x^2}{0.1} = 10^{-4.8}$$

$$\therefore x^2 = 10^{-5.8} \implies x = 10^{-2.9}$$

$$\therefore [H_3O^+] = 10^{-2.9} \text{ mol dm}^{-3} \text{ and } pH = 2.9$$

Bases in water. The pH of a solution of a weak base in water is calculated in a similar manner to that of a weak acid. Instead of finding $[H_3O^+]$ from a given K_a, $[OH^-]$ is found from a given K_b. For example, the pH of 0.5 mol dm^{-3} ammonia is calculated as follows.

$$NH_3(aq) + H_2O(l) \rightleftharpoons NH_4^+(aq) + OH^-(aq)$$

	NH$_3$	NH$_4^+$	OH$^-$	
initial amount/mol	0.5	0	0	in 1 dm^3

let x mol dm^{-3} of OH$^-$(aq) be present at equilibrium

equilibrium amount/mol	$(0.5 - x)$	x	x	in 1 dm^3

Since $K_b = \frac{[NH_4^+][OH^-]}{[NH_3]} = 10^{-4.8}$ mol dm^{-3} and, assuming $[NH_3]_{initial} \approx [NH_3]_{equilibrium}$,

$$\therefore \frac{x^2}{0.5} = 10^{-4.8} \implies x = 10^{-2.75}$$

$$\therefore [OH^-] = 10^{-2.75} \text{ mol dm}^{-3} \implies [H_3O^+] = 10^{-11.25} \text{ mol dm}^{-3}$$

$$\therefore pH = 11.25 \qquad (K_w = [H_3O^+][OH^-])$$

Buffer solutions

> A buffer solution tends to resist a change in pH when small amounts of acid or base are added to it.

There are two types of buffer solution:

1 an acidic buffer, pH < 7, made by mixing a weak acid and its conjugate base in roughly equimolar proportions;
2 a basic buffer, pH > 7, made by mixing a weak base and its conjugate acid in roughly equimolar proportions.

Acidic buffer. For example, a solution that contains 0.1 mole of ethanoic acid and 0.1 mole of sodium ethanoate in 1 dm^3 is an acidic buffer solution. Since ethanoic acid is a weak acid and is hardly dissociated at all, whereas sodium ethanoate is fully dissociated, the following assumptions can be made.

$[HA]_{initial} \approx [HA]_{equilibrium}$ and $[NaA]_{initial} \approx [A^-]_{equilibrium}$

$$HA(aq) + H_2O(l) \rightleftharpoons H_3O^+(aq) + A^-(aq)$$

equilibrium amount/mol   0.1   0.1   in 1 dm^3

Since $K_a = \dfrac{[H_3O^+][A^-]}{[HA]}$ and the ratio $\dfrac{[A^-]}{[HA]} = 1$ (from above)

$\therefore K_a = [H_3O^+]$ or $pK_a = pH$

$\therefore pH = 4.8$ ($K_a = 10^{-4.8}$ mol dm^{-3})

The table below summarizes the effects of adding a little acid or base to the buffer solution.

addition of acid	addition of base
A^- ions are protonated $\therefore [A^-]$ decreases a little $\therefore [HA]$ increases a little but $[A^-]/[HA]$ is still close to unity (as above) $\therefore pH \approx pK_a = 4.8$	HA molecules are deprotonated $\therefore [HA]$ decreases a little $\therefore [A^-]$ increases a little but $[A^-]/[HA]$ is still close to unity (as above) $\therefore pH \approx pK_a = 4.8$

Basic buffer. For example, a solution that contains 0.1 mole of ammonia and 0.1 mole of ammonium chloride in 1 dm^3 is a basic buffer solution. Since ammonia is a weak base and is hardly dissociated at all, whereas ammonium chloride is fully dissociated, the same assumptions can be made again.

i.e. $[NH_3]_{initial} \approx [NH_3]_{equilibrium}$; $[NH_4Cl]_{initial} \approx [NH_4^+]_{equilibrium}$

Using the same argument as that used for ethanoic acid and ethanoate ions (given K_b for ammonia $= 10^{-4.8}$ mol dm^{-3} and $K_w = 10^{-14}$ mol^2 dm^{-6}),

$K_b = \dfrac{[NH_4^+][OH^-]}{[NH_3]}$ but $\dfrac{[NH_4^+]}{[NH_3]} = 1$

$\therefore K_b = [OH^-] = K_w/[H_3O^+]$ ($K_w = [H_3O^+][OH^-]$)

$\therefore [H_3O^+] = K_w/K_b$ or $pH = pK_w - pK_b$

$\therefore pH = 9.2$

Since the ratio $[NH_4^+]/[NH_3]$ remains close to unity on addition of small amounts of acid or base, the pH remains close to the value of $(pK_w - pK_b)$.

Measuring pH

Indicators. An indicator is a compound whose colour in solution depends on the pH of the particular solution. It provides a chemical means of estimating pH by the shade of colour adopted. The dependence of colour on pH is a function of the weak acidity of the indicator itself; a typical indicator is a weak organic acid whose conjugate base is a different colour from that of the acid.

$$HIn(aq) + H_2O(l) \rightleftharpoons H_3O^+(aq) + In^-(aq)$$

colour A 　　　　　　　　　　　colour B

at low pH	at high pH
$[H_3O^+]$ is high ∴ the position of the above equilibrium lies to the left ∴ colour A is seen.	$[H_3O^+]$ is low ∴ the position of the above equilibrium lies to the right ∴ colour B is seen.

The change in colour takes place when $[HIn] = [In^-]$.
From the equilibrium law,

$$K_{In} = \frac{[H_3O^+][In^-]}{[HIn]}$$

when $[In^-] = [HIn]$, $K_{In} = [H_3O^+]$ or $pH = pK_{In}$. So an indicator is changing colour at a pH equal to its pK_{In}. Two common indicators are illustrated in the following table.

indicator	colour A	colour B	K_{In} / mol dm^{-3}	pH range of colour change
phenolphthalein	colourless	pink	$10^{-9.4}$	8.3–10.0
methyl orange	red	yellow	$10^{-3.3}$	3.1–4.4

pH meter. A physical method for measuring pH involves the use of a glass electrode. This electrode has a glass membrane of unusually low conductivity for glass. The inside of the membrane is in contact with a solution of standard acidity. The outside of the membrane is put in contact with the solution whose pH is to be measured.

The different concentrations of H_3O^+ in contact with the inside and outside of the membrane cause a small potential difference between the surfaces. This potential difference is directly proportional to the pH of the unknown solution. It is measured by comparing it with that of a standard reference electrode built into the glass electrode. The readings recorded by a valve voltmeter can be calibrated directly into pH units if the electrode is dipped initially into a buffer solution of known pH.

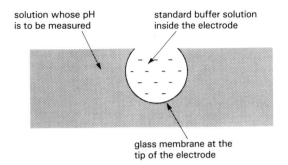

Acid-base titrations. In the titration of a base against added acid, the equivalence-point is reached when the molar ratio of base to acid is the same as the ratio given by the equation for the reaction. Under these conditions, the solution contains only a salt and water, and so the addition of any more acid from the burette is bound to cause a much larger change in pH than occurs when the solution still contains some unreacted base.

The equivalence-point of a titration can, therefore, be observed by adding an indicator whose pK_{In} is approximately equal to the pH of the salt solution in question. Up to the equivalence-point, the indicator shows one colour, but the rapid pH change at the equivalence-point causes it to change to the other colour. For a particular titration, the choice of a suitable indicator is governed by the pH of the salt produced. On page 86, the hydrolysis of salts is discussed. For example, the salt of a weak base and a strong acid (such as ammonium chloride) is acidic, whereas the salt of a strong base and a weak acid (such as sodium ethanoate) is alkaline. This is illustrated by the curves below and over the page.

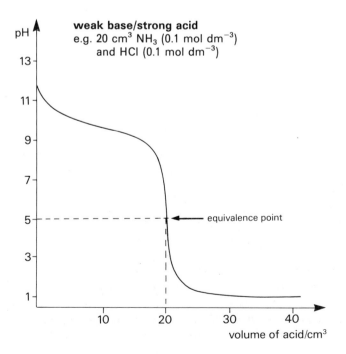

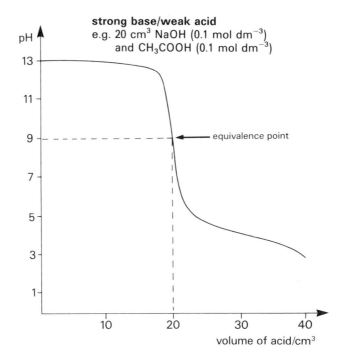

Methyl orange would indicate the first equivalence-point but not the second: phenolphthalein would indicate the second but not the first. The data for the above titration curves are obtained by using a pH meter to measure the pH after small additions of acid to a volume of base.

6.3 ELECTRON TRANSFER

Redox potentials

There are two opposing processes that tend to occur when a metal electrode is dipped into a solution of its own ions

1. The metal atoms tend to lose electrons and form hydrated cations in solution. The outer-shell electrons are left within the lattice which, therefore, becomes negatively-charged: $M(s) \rightarrow M^{n+}(aq) + n\,e^-$.
2. The hydrated cations in the solution tend to withdraw electrons from the electrode to form metal atoms. They make up an extension of the lattice which, therefore, becomes positively charged: $M^{n+}(aq) + n\,e^- \rightarrow M(s)$.

The two opposing processes are illustrated on the facing page for a zinc electrode dipping into a solution of $Zn^{2+}(aq)$ ions (a zinc 'half-cell').

One or other of these tendencies dominates in a given half-cell, so the potential of the electrode becomes different from that of the solution. As the potential is set up, however, the balance between the two tendencies changes. For example, a negatively charged electrode attracts cations, and thus the conversion of cations to atoms becomes

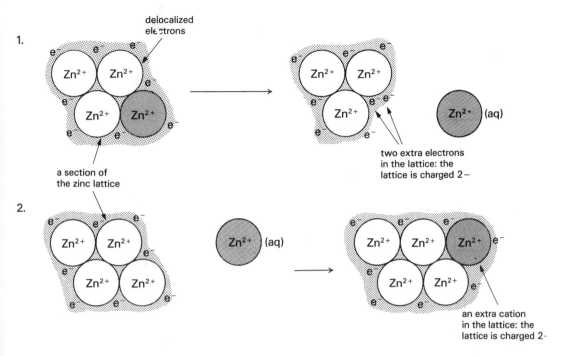

more likely. However, the further loss of cations from the lattice, which is the process responsible for making the electrode negative in the first place, becomes less likely.

The potential difference, therefore, builds up until neither one nor the other tendency is dominant. Under these conditions, the electrode is in equilibrium with the solution and the potential difference between the electrode and the solution is called the electrode (or redox) potential, E.

$$M^{n+}(aq) + n\,e^- \rightleftharpoons M(s) \quad \text{p.d.} = E \text{ volts}$$

Standard electrode (or redox) potential, $E^{\ominus}$

> The standard electrode potential ($E^{\ominus}$) of a half-cell is the potential difference between the half-cell and a standard hydrogen half-cell; all ionic concentrations are 1 mol dm^{-3} and measurements are made at 298 K and 0.1 MPa pressure

It is impossible to obtain an absolute measurement of an electrode potential, E. In making an electrical contact to the solution, a new electrode potential is set up between the contact and the solution. This obscures the value of the potential to be measured. The problem of measuring an electrode potential is entirely similar to that of measuring height or enthalpy (see page 62). Just as with enthalpy a standard enthalpy is defined so that the enthalpy of any other system can be compared with the standard enthalpy, so for electrode processes a standard half-cell is chosen: the hydrogen half-cell. All other electrode potentials are compared with the potential of this standard half-cell. For example, the standard electrode potential, $E^{\ominus}$ of iron is determined as shown overleaf.

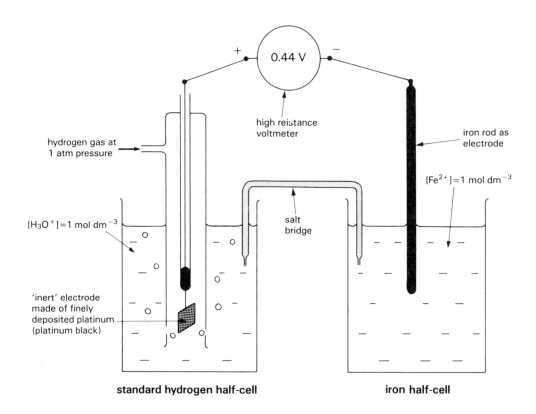

standard hydrogen half-cell   **iron half-cell**

The standard half-cell contains an 'inert' electrode made of platinum black. This is needed to bring hydrogen molecules and ions into contact at its surface, while the platinum atoms show almost no tendency to go into solution. The two half-cells are connected by a salt bridge which contains saturated potasssium chloride solution. The salt bridge ensures electrical contact without introducing another electrode into the system. The reading on the voltmeter gives $E^{\ominus}$ from the iron(II)/iron half-cell.

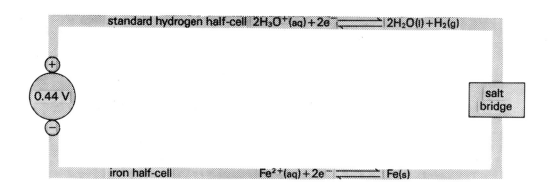

Inert electrodes. To determine $E^\ominus$ for the equilibrium between two aqueous ions, a second inert electrode becomes necessary. For example, $E^\ominus$ for the equilibrium between iron(III) and iron(II) ions is measured using an inert platinum electrode as shown below.

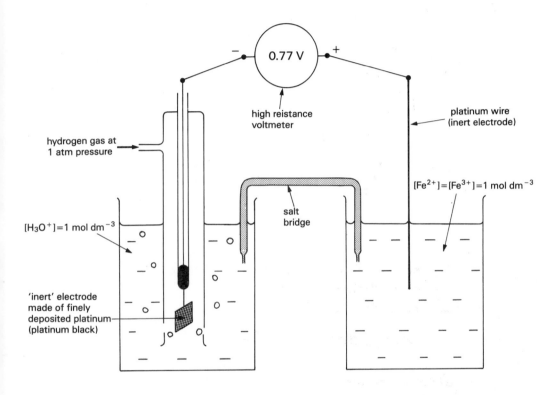

The reading on the voltmeter gives $E^\ominus$ for the iron(III)/iron(II) half-cell.

$$Fe^{3+}(aq) + e^- \rightleftharpoons Fe^{2+}(aq) \qquad E^\ominus = +0.77 \text{ V}$$

Reduction and oxidation

> Reduction is a process of electron gain; oxidation is a process of electron loss. A redox process is one in which electron transfer takes place.

An atom loses or gains control of electrons simply as a result of being bonded to other atoms. The oxidation number of an element in a compound shows the number of electrons over which each atom has lost or gained control during the formation of the compound. For example, the oxidation numbers of iron and oxygen in different compounds are illustrated overleaf.

	iron		oxygen	
Fe(II)	Fe²⁺	[S²⁻ Lewis structure]	O(−II)	[H₂O with δ+ on H, δ− on O]
	Each iron atom has lost control of two electrons to each sulphur atom		Each oxygen atom has gained partial control of two electrons, one from each hydrogen atom	
Fe(III)	Fe³⁺	([Cl⁻])₃	O(−I)	[H₂O₂ structure with δ+ on H, δ− on O]
	Each iron atom has lost control of three electrons (to three separate chlorine atoms)		Each oxygen atom has gained partial control of one electron (from the hydrogen atom bonded to it)	

The use of oxidation numbers makes it easier to interpret redox reactions. During oxidation, the oxidation number of an element increases; during reduction, the oxidation number of an element decreases. For example,

1 Iron(II) chloride is oxidized by chlorine to iron(III) chloride.

$$2\overset{II}{Fe}Cl_2(s) + \overset{0}{Cl}_2(g) \xrightarrow{heat} 2\overset{III-I}{Fe}Cl_3(s)$$

Electrons are lost from iron: $Fe^{2+} \rightarrow Fe^{3+} + e^-$. The oxidation number of iron increases from two to three.

2 Manganese(IV) oxide is reduced to manganese(II) chloride by concentrated hydrochloric acid.

$$\overset{IV}{Mn}O_2(s) + 4H\overset{-I}{Cl}(conc) \longrightarrow \overset{II}{Mn}Cl_2(s) + 2H_2O(l) + \overset{0}{Cl}_2(g)$$

Electrons are gained by manganese: $Mn(IV) + 2e^- \rightarrow Mn(II)$.
The oxidation number of manganese decreases from four to two.

A substance that causes oxidation is called an oxidizing agent (or oxidant); a substance that causes reduction is called a reducing agent (or reductant). In the above two examples, the oxidants and reductants are as follows.

oxidant	reductant
chlorine	iron(II) chloride
manganese(IV) oxide	hydrochloric acid

In some cases, a substance may undergo simultaneous oxidation and reduction without the intervention of any other substance. This process is called disproportionation. For example, in aqueous solution, copper(I) salts disproportionate to copper and copper(II) salts.

$$Cu^+(aq) \longrightarrow Cu^{2+}(aq) + e^- \quad \text{while} \quad Cu^+(aq) + e^- \longrightarrow Cu(s)$$

$$\therefore 2\overset{I}{Cu}{}^+(aq) \longrightarrow \overset{0}{Cu}(s) + \overset{II}{Cu}{}^{2+}(aq)$$

In the same way that an acid has a conjugate base produced by proton transfer (page 85), so an oxidant has a conjugate reductant produced by electron transfer. A strong oxidant has a weak conjugate reductant and vice versa. For example,

	oxidant + ne −	⇌	reductant	
strength as oxidant ↓	$F_2(g)$ $MnO_4^-(aq)$ $Cl_2(g)$ $Fe^{3+}(aq)$ $O_2(g)$ $Cu^{2+}(aq)$ $HSO_4^-(aq)$ $Zn^{2+}(aq)$ $Mg^{2+}(aq)$ $K^+(aq)$	⇌ ⇌ ⇌ ⇌ ⇌ ⇌ ⇌ ⇌ ⇌ ⇌	$2F^-(aq)$ $Mn^{2+}(aq)$ $2Cl^-(aq)$ $Fe^{2+}(aq)$ $H_2O_2(aq)$ $Cu(s)$ $SO_2(g)$ $Zn(s)$ $Mg(s)$ $K(s)$	strength as reductant ↓

Using $E^\ominus$ data to predict reactivity

The general equation for a half-cell at equilibrium can be written in the following form:

$$\text{oxidant} + ne^- \rightleftharpoons \text{reductant}$$

The value of $E^\ominus$ for the half-cell is a measure of the relative oxidizing or reducing properties of the couple. For example, in the following two half-cells,

$$MnO_4^-(aq) + 8H_3O^+(aq) + 5e^- \rightleftharpoons Mn^{2+}(aq) + 12H_2O(l)$$

$$Cl_2(g) + 2e^- \rightleftharpoons 2Cl^-(aq)$$

the $E^\ominus$ data make it possible to predict whether chlorine is likely to oxidize a manganese(II) salt, or whether manganate(VII) is likely to oxidize a chloride salt.

1 Consider a cell made up from the two half-cells concerned;

Mn(VII)/Mn(II) $E^\ominus = +1.52$ V
Cl(0)/Cl(−I) $E^\ominus = +1.36$ V

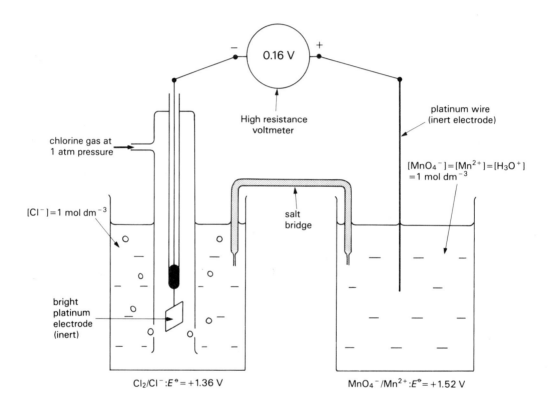

Cl_2/Cl^- : $E^\ominus = +1.36$ V MnO_4^-/Mn^{2+} : $E^\ominus = +1.52$ V

2 In this case, the manganese half-cell is positive (by 0.16 volt).
3 Electrons tend to flow through the wire from negative to positive: hence, the chlorine half-cell must produce electrons for the manganese half-cell to use up.

$10Cl^-(aq) \longrightarrow 5Cl_2(g) + 10e^-$

$2MnO_4^-(aq) + 16H_3O^+(aq) + 10e^- \longrightarrow 2Mn^{2+}(aq) + 24H_2O(l)$

4 In other words, manganate(VII) tends to oxidize chloride ions to chlorine.

If the 'half-equations' for various half-cells are listed in order of decreasing $E^\ominus$, the following generalization can be made: an oxidant tends to oxidize any reductant below it on the list. This generalization can be checked by the procedure used above for the chlorine and manganese half-cells.

For example, fluorine tends to oxidize any reductant in the list, whereas a potassium salt is unlikely to oxidize any of the reductants listed on the next page.

oxidant + ne^-	⇌ reductant	$E^\ominus$/volt
$F_2(g) + 2e^-$	⇌ $2F^-(aq)$	+2.87
$MnO_4^-(aq) + 8H_3O^+ + 5e^-$	⇌ $Mn^{2+}(aq) + 12H_2O(l)$	+1.52
$Cl_2(g) + 2e^-$	⇌ $2Cl^-(aq)$	+1.36
$Fe^{3+}(aq) + e^-$	⇌ $Fe^{2+}(aq)$	+0.77
$O_2(g) + 2H_3O^+(aq) + 2e^-$	⇌ $H_2O_2(aq) + 2H_2O(l)$	+0.68
$Cu^{2+}(aq) + 2e^-$	⇌ $Cu(s)$	+0.34
$HSO_4^-(aq) + 3H_3O^+(aq) + 2e^-$	⇌ $SO_2(g) + 5H_2O(l)$	+0.17
$Zn^{2+}(aq) + 2e^-$	⇌ $Zn(s)$	−0.76
$Mg^{2+}(aq) + 2e^-$	⇌ $Mg(s)$	−2.38
$K^+(aq) + e^-$	⇌ $K(s)$	−2.92

Precautions in prediction. When using $E^\ominus$ data to predict redox changes, there are two major precautions to bear in mind.

1. The predictions are made solely from the viewpoint of energy. Although a process may be energetically favourable, the activation energy (page 73) is often so high that its rate is infinitely slow under the reaction conditions.
2. The reaction conditions of many redox processes are far removed from the standard conditions to which $E^\ominus$ data refer. Under non-aqueous conditions, the use of $E^\ominus$ data to predict and interpret redox change is very limited. In cases where the concentrations of the substances are varied from 1 mol dm^{-3}, the equilibrium ideas discussed on page 82 are applicable.

7 Electrochemistry

7.1 CONDUCTIVITY

Electrolytes

> An electrolyte is a substance that conducts electricity in the molten or aqueous phase as a result of the movement of ions through the system.

The ions transport charge to the electrodes where electrolysis takes place: electrons flow through the wire from anode to cathode. For example, molten lead(II) bromide conducts electricity but is decomposed to lead and bromine in the process.

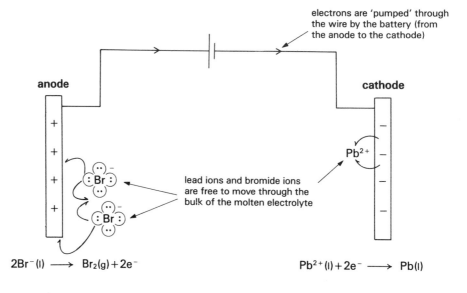

In an aqueous solution of an electrolyte, the ability of the ions to transport charge is controlled by their mobility, and is dependent on the number of ions between the electrodes. The second of these two factors depends on the concentration and degree of dissociation of the electrolyte concerned.

The degree of dissociation, α, of an electrolyte is the fraction of each mole of electrolyte that is dissociated into ions in a solution of the electrolyte at equilibrium.

For example, in an aqueous solution of an electrolyte A^+B^- of concentration c mol dm^{-3}, the equilibrium dissociation constant, K is defined as follows.

	$A^+B^-(aq)$	$\rightleftharpoons$	$A^+(aq)$	$+ B^-(aq)$	
initial amount/mol	c		0	0	in 1 dm^3

At equilibrium, $(\alpha \times c)$ mol of electrolyte is dissociated,

equilibrium amount/mol	$(c - \alpha c)$		αc	αc	in 1 dm^3

By the equilibrium law,

$$K = \frac{[A^+][B^-]}{[AB]} = \frac{(\alpha c)^2}{c - \alpha c} = \frac{\alpha^2 c^2}{c(1 - \alpha)}$$

$$\Rightarrow K = \frac{\alpha^2 c}{1 - \alpha} \quad \text{the Ostwald dilution law}$$

In cases where the electrolyte is weak and, therefore, hardly dissociated at all, α is very small. Under these conditions, the Ostwald dilution law can be simplified because $(1 - \alpha) \approx 1$ to a good approximation.

$$\therefore K = \alpha^2 c \quad \text{or} \quad \alpha = \sqrt{K/c}$$

For example, ethanoic acid is a weak electrolyte whose dissociation constant is $10^{-4.8}$ mol dm^{-3} (see the first example on page 88). The degree of dissociation of a solution containing 0.1 mol dm^{-3} of the acid is readily found.

$$\alpha = \sqrt{\frac{10^{-4.8}}{10^{-1}}} = (10^{-3.8})^{1/2} = 10^{-1.9} = 0.013$$

$$\therefore \alpha = 1.3\%$$

Trichloroethanoic acid CCl_3COOH ($K_a = 10^{-0.65}$ mol dm^{-3}) is a stronger acid than ordinary ethanoic acid (see page 219), and therefore has a higher degree of dissociation. For a 0.1 mol dm^{-3} solution, the Ostwald dilution law is applied as follows:

$$\frac{\alpha^2 \times 0.1}{1 - \alpha} = 10^{-0.65} \Rightarrow \alpha = 74.5\%$$

Conductivity, κ, and molar conductivity Λ

The conductivity, κ, of a solution is determined by measuring the resistance, R, of the solution in a cell whose electrodes have a cross-sectional area, A, and are a distance, l, apart; $\kappa = l/RA$.

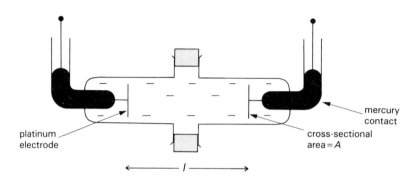

The resistance of the solution in the cell is measured using a Wheatstone bridge circuit as shown below.

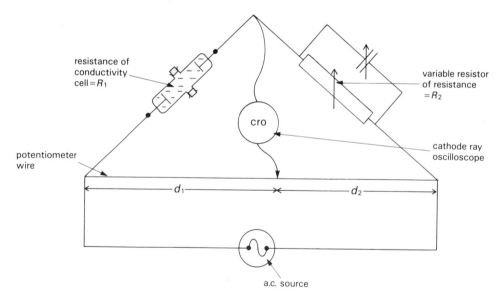

There are four different resistances being balanced by the bridge circuit: the cell R_1, the resistor setting R_2 and the two separate lengths of potentiometer wire, d_1 and d_2. The contact is slid up and down the potentiometer wire until the oscilloscope (cro) shows a stationary loop on its screen. This means that the bridge is balanced, and the following relationship holds.

$$\frac{R_1}{R_2} = \frac{d_1}{d_2}$$

It is important to use an a.c. source to prevent electrolysis from happening at the electrodes. In most cases, the cell dimensions are not known precisely, and the so the cell constant, l/A is found by a preliminary experiment in which the resistance of a solution of *known* conductivity is measured. The most common standard is 0.1 mol dm^{-3} potassium chloride solution whose conductivity at 298 K is 1.29 Ω^{-1} m^{-1}. Using the above apparatus, a table of values of κ at different electrolyte concentrations, c, can be obtained. The results for a typical strong electrolyte are shown on the graph below.

At low concentrations, κ is small because there are only a few ions between the electrodes; at high concentrations, κ becomes small again because the ions tend to associate and hinder one another's progress to the electrodes. In order to obtain a conductivity measure that is independent of these factors, a molar conductivity, Λ, is defined as follows.

> The molar conductivity, Λ, of a solution is the conductivity of a 'molar volume' of the solution: this is the volume that contains exactly one mole of electrolyte between two electrodes 1 metre apart at the specified temperature.

For example, the relationship between the conductivity, κ, and the molar conductivity, Λ, of 0.002 mol dm^{-3} potassium chloride is illustrated overleaf.

conductivity = κ	conductivity = Λ
The conductivity of 1 m³ of solution between electrodes 1 m apart is κ	There are 0.002 moles of KCl in 1 dm³ ∴ There are 2 moles in 1000 dm³ = 1 m³ ∴ 1 mole of KCl is present in 0.5 m³

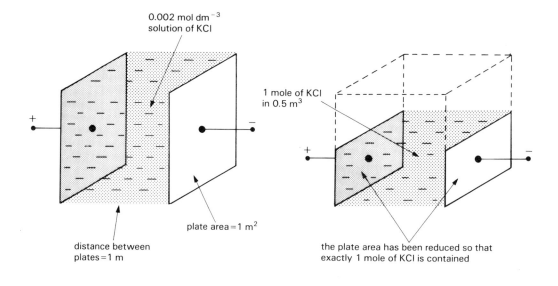

The general relationship is $\Lambda = \kappa V$, where V is the volume of the solution (in cubic metres) that contains one mole of electrolyte. Since κ is measured in $\Omega^{-1} m^{-1}$ and V is in $m^3 \, mol^{-1}$, the units of Λ are $\Omega^{-1} m^2 mol^{-1}$.

Infinite dilution. The molar conductivity of an electrolyte at 'infinite' dilution has a particular significance. Under these conditions, all electrolytes must be fully dissociated and so the value of Λ reaches a maximum. This value is given the symbol Λ_∞ and it follows from the definition that, if Λ_c is the molar conductivity of a solution of concentration c, the degree of dissociation of the electrolyte is given by:

$$\alpha = \frac{\Lambda_c}{\Lambda_\infty}$$

Λ_∞ is determined for a strong electrolyte as follows.

    1 κ is measured for various solutions of different known concentrations, c.
    2 Λ_c is calculated from each value of κ and c.
    3 A plot of Λ_c against $\sqrt{c}$ is carried out and extrapolated back to $c = 0$ (infinite dilution).

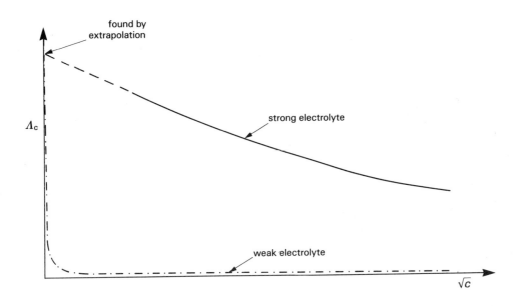

For a weak electrolyte, the above method fails because the molar conductivity increases too slowly with increasing dilution. This is shown on the graph below.

Λ_∞ is determined for a weak electrolyte by applying Kohlrausch's law of ionic mobilities.

Kohlrausch noticed that there was a constant difference between the molar conductivities of various sodium and potassium salts (strong electrolytes). For example, the figures at 298 K for the chlorides, nitrates and sulphates of the two metals are shown below in $\Omega^{-1} m^2 mol^{-1}$.

$\Lambda_\infty / \Omega^{-1} m^2 mol^{-1}$	chloride	nitrate	sulphate
potassium	0.01499	0.01449	0.02331
sodium	0.01265	0.01215	0.02097
difference	0.00234	0.00234	0.00234

This led him to propose that the molar conductivity of an electrolyte at infinite dilution is the sum of two independent ionic mobilities: the molar conductivity of the cation and that of the anion at infinite dilution.

> Kohlrausch's law: for $AB_{(aq)} \rightleftharpoons A^+_{(aq)} + B^-_{(aq)}$
>
> $\Lambda_\infty = \lambda_\infty(A^+) + \lambda_\infty(B^-)$
>
> where λ_∞ is the molar conductivity of an ion at infinite dilution (the mobility of the ion).

Kohlrausch's 'constant differences' now reduce to

$$[\lambda_\infty(K^+) - \lambda_\infty(Na^+)] = 0.00234 \; \Omega^{-1} \; m^2 \; mol^{-1}$$

The law of ionic mobilities is applied as follows to calculate Λ_∞ for a weak electrolyte like ethanoic acid (H^+, CH_3COO^- shown below as H^+, A^-).

1. Measure Λ_∞ for two strong electrolytes which between them contain the same ions found in aqueous ethanoic acid; for example, hydrochloric acid and sodium ethanoate.
2. Measure Λ_∞ for the strong electrolyte that contains the additional ions from the first two experiments. In this case it is sodium chloride.
3. Apply Kohlrausch's law to each Λ_∞ and derive an expression for Λ_∞ of ethanoic acid.

electrolyte	Kohlrausch's law	$\Lambda_\infty / \Omega^{-1} \; m^2 \; mol^{-1}$
(i) HCl	$\lambda_\infty(H^+) + \lambda_\infty(Cl^-)$	0.04262
(ii) NaA	$\lambda_\infty(Na^+) + \lambda_\infty(A^-)$	0.00910
(iii) NaCl	$\lambda_\infty(Na^+) + \lambda_\infty(Cl^-)$	0.01265

Λ_∞ of HA = $\lambda_\infty(H^+) + \lambda_\infty(A^-)$ = (i) + (ii) − (iii)

∴ Λ_∞ of CH_3COOH = 0.03907 $\Omega^{-1} \; m^2 \; mol^{-1}$

7.2 ELECTROLYSIS

Selective discharge

Electrolysis occurs at the electrodes dipping into a molten or aqueous electrolyte, when a current passes. Cations are discharged at the cathode by gaining electrons; anions are discharged at the anode by losing electrons. If the system contains a mixture of electrolytes, one ion tends to be discharged in preference to the others. This process is called selective discharge.

The order in which ions are discharged can be predicted from redox potential data (pages 97 and 99). The ions that are the most readily formed are the least readily discharged. For example, in the manufacture of sodium, molten sodium chloride is electrolysed in the presence of 60% calcium chloride (added to lower salt's melting point). Sodium is selectively discharged at the cathode in agreement with the redox data.

$Na^+(aq) + e^- \rightleftharpoons Na(s) \quad E^\ominus = -2.71 \; V$

$Ca^{2+}(aq) + 2e^- \rightleftharpoons Ca(s) \quad E^\ominus = -2.87 \; V$

Aqueous electrolytes. The electrolysis of an aqueous solution of an ionic salt often produces hydrogen and oxygen at the electrodes instead of metallic and other non-metallic products. For example, if brine is electrolysed between graphite electrodes, little chlorine and no sodium is produced. The reasons for this are illustrated on page 107.

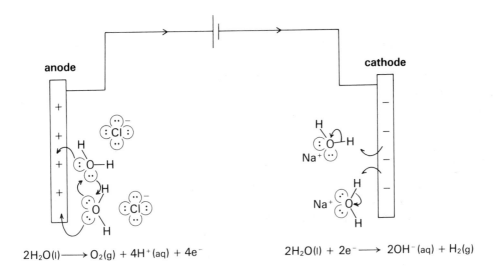

The ions arrive at the electrodes surrounded by water molecules. Under these conditions, there is a competition between the ions and their solvating molecules for electron transfer at the electrodes. In the above case, the relevant redox data are shown below.

cathode	$E^\ominus/V$	anode	$E^\ominus/V$
Na^+/Na	−2.71	$Cl_2/2Cl^-$	+1.36
H_2O/H_2	−0.40	$O_2/2H_2O$	+1.23

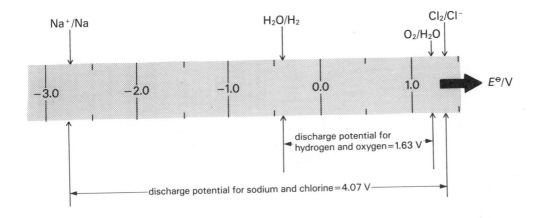

These values suggest that the discharge of hydrogen and oxygen is favoured over that of sodium and chlorine.

Overvoltage

In certain circumstances, the order of discharge of aqueous ions does not follow the predictions of redox data. For example, sodium and chlorine can be obtained from brine providing a mercury cathode and a steel anode are used. The change in selective discharge is brought about by three factors.

1 Hydrogen has a high *overvoltage* on mercury.
2 Sodium forms an amalgam with mercury which effectively lowers its discharge potential.
3 Oxygen has a high *overvoltage* on steel.

An overvoltage is the additional potential (in excess of the discharge potential) required to discharge an ion at the electrode. It is a measure of the activation energy of the electrode process and is only a substantial factor in the formation of gases at a solid surface.

8 The Chemistry of the Metals

8.1 GROUP I: THE ALKALI METALS

Structure

A Group I metal atom has one outer-shell electron (ns^1) and an empty $(n-1)$d subshell. The degree of control of the outer-shell electrons is the lowest for any group: the first ionization energy of each of the Group I elements represents a trough on the diagram below.

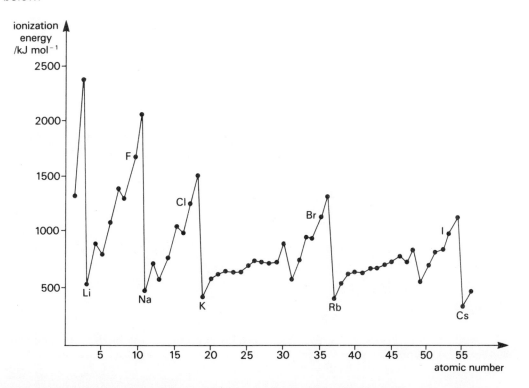

The elements owe their metallic character to this low degree of control of outer-shell electrons (see page 23). All the Group I metals adopt a body-centred cubic lattice of single charged cations in a cloud of delocalized outer-shell electrons.

1.2 Compound formation

The metals react vigorously with most non-metals to form compounds with a high percentage of ionic character. Ionic bonding is favoured because the Group I cations are comparatively large and are only singly charged (Fajans' rules, see page 16). The low ionization energy required to form the cations is more than compensated by the energy released when the ionic lattice forms. The only stable oxidation state for the metals in their compounds is +I. The first electron is easier to remove than any others because it is well shielded from the nucleus and the core charge (effective nuclear charge) is only one; a core charge equals the charge of the nucleus less that of the shielding inner-shell electrons.

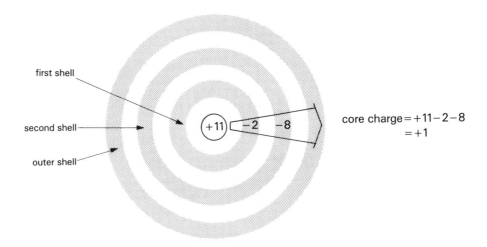

To remove a second electron is much more difficult. The electron must come from a closer innershell where the core charge is much greater.

Acid–base properties

The metals are called alkali metals for two reasons:

1 they react with water to give alkaline solutions;
2 their oxides and hydroxides are soluble (unlike most other metal oxides and hydroxides). These dissolve to give alkaline solutions.

$$2Na(s) + 2H_2O(l) \longrightarrow 2Na^+(aq) + 2OH^-(aq) + H_2(g)$$
$$Na_2O(s) + H_2O(l) \longrightarrow 2Na^+(aq) + 2OH^-(aq)$$

Group I metal cations are the least hydrolysed in solution (see page 87); the polarizing power of the cations is the smallest as a result of their large size and low charge.

Redox properties

The elements are all strong reductants showing a marked tendency to lose electrons. The standard redox potentials of the metals are high and negative. For example,

$$Na^+(aq) + e^- \rightleftharpoons Na(s) \quad E^\ominus = -2.71 \text{ V}$$
$$K^+(aq) + e^- \rightleftharpoons K(s) \quad E^\ominus = -2.92 \text{ V}$$

Most non-metals and many of their compounds are reduced by a Group I metal. For example, oxygen, chlorine and water are all reduced by sodium.

$$4\overset{0}{Na}(s) + \overset{0}{O_2}(g) \longrightarrow 2\overset{I}{Na_2}\overset{-II}{O}(s)$$

$$2\overset{0}{Na}(s) + \overset{0}{Cl_2}(g) \longrightarrow 2\overset{I}{Na}\overset{-I}{Cl}(s)$$

$$2\overset{0}{Na}(s) + 2\overset{I}{H_2}O(l) \longrightarrow 2\overset{I}{Na}{}^+(aq) + 2OH^-(aq) + \overset{0}{H_2}(g)$$

The reaction between oxygen and an alkali metal leads to the production of a number of different oxides. The metal lattice can lose four, two or sometimes, only one electron to each oxygen molecule. This is illustrated in the following table.

number of electrons accepted per O_2 molecule	structure	example
one		K^+ O_2^- potassium superoxide, KO_2
two		$2Na^+$ O_2^{2-} sodium peroxide, Na_2O_2
four		$4Li^+$ $2O^{2-}$ lithium oxide, Li_2O

Superoxides are stable only when the lattice contains cations of a very low polarizing power e.g. K^+, Rb^+. They are formed when the metal is burnt in excess oxygen.

The reducing properties of sodium are used in the manufacture of titanium from its chloride. Sodium displaces titanium from titanium(IV) chloride.

$$4\overset{0}{Na} + \overset{IV}{Ti}Cl_4 \longrightarrow 4\overset{I}{Na}Cl + \overset{0}{Ti}$$

8.2 GROUP II: THE ALKALINE EARTH METALS

Similarities to Group I

The elements are all reactive metals whose atoms have a low degree of control of their outer-shell electrons. Each atom has one more outer-shell electron and one more proton than the corresponding Group I metal atom, and so the stable oxidation state and core charge are not the same as those of Group I. However, many similarities in properties exist.

1. The Group II compounds are largely ionic in character.
2. The elements react with water to give alkaline suspensions of partially soluble metal hydroxides.

$$Ca(s) + 2H_2O(l) \longrightarrow Ca(OH)_2(s) + H_2(g)$$

$$Ca(OH)_2(s) \xrightleftharpoons{(aq)} Ca^{2+}(aq) + 2OH^-(aq)$$

3. The Group II metals are all strong reductants and reduce non-metals, water and the salts of less reactive metals. Their redox potentials are large and negative. For example,

$$Mg^{2+}(aq) + 2e^- \rightleftharpoons Mg(s) \quad E^\ominus = -2.38 \text{ V}$$

$$\overset{0}{Mg}(s) + \overset{0}{Cl_2}(g) \longrightarrow \overset{II\ -I}{MgCl_2}(s)$$

$$\overset{0}{Mg}(s) + \overset{II}{Zn}^{2+}(aq) \longrightarrow \overset{II}{Mg}^{2+}(aq) + \overset{0}{Zn}(s)$$

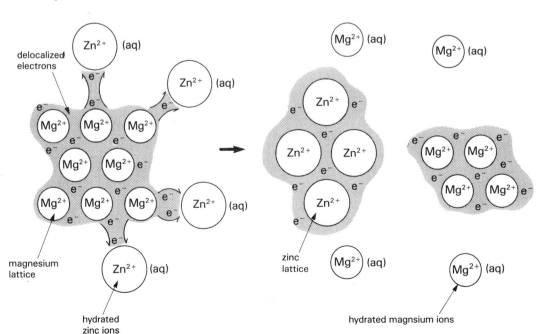

$$Mg(s) + Zn^{2+}(aq) \longrightarrow Mg^{2+}(aq) + Zn(s)$$

The last example is an example of a *displacement reaction*: a more reactive metal displaces a less reactive one from a solution containing the latter's ions.

Differences from Group I

The major differences all stem from the formation of a doubly charged cation instead of a singly charged one. More energy is required for this process (IE in the diagram below), but there is also a large increase in lattice energy (U) when the lattice of a Group II compound forms. Compare the formation of the chlorides shown below (see page 65).

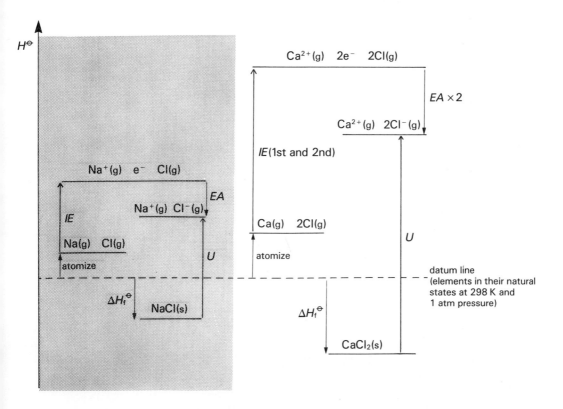

The higher charge and smaller size of Group II cations give them greater polarizing powers (Fajans' rules, page 16). The following properties can be explained in terms of the higher polarizing power of the alkaline earth metal cations.

1. The insolubility of Group II oxides, hydroxides and carbonates (the corresponding Group I compounds are soluble): there is a larger degree of covalency in a Group II lattice, and this makes dissociation less likely.
2. The greater degree of hydrolysis of the Group II cations in aqueous solution: the ions polarize the water ligands more extensively than do the larger, singly charged Group I cations. This is illustrated overleaf.

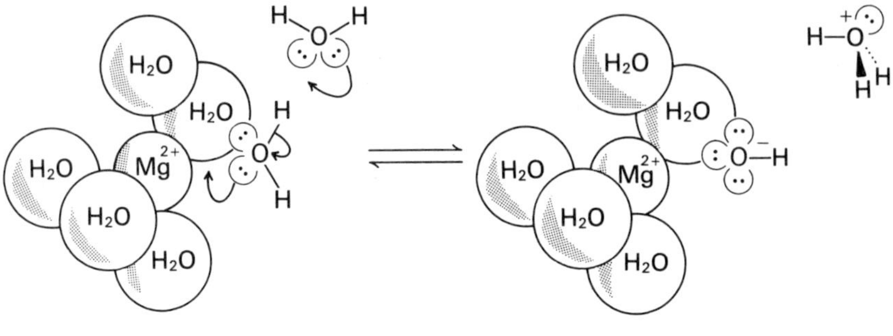

Compare the pH of 1 mol dm^{-3} Na$^+$(aq) = 7.0

with the pH of 1 mol dm^{-3} Mg^{2+}(aq) = 6.9.

3 The thermal instability of certain Group II compounds (for example, the carbonates and hydroxides): the high polarizing power makes the formation of an oxide lattice likely. For example,

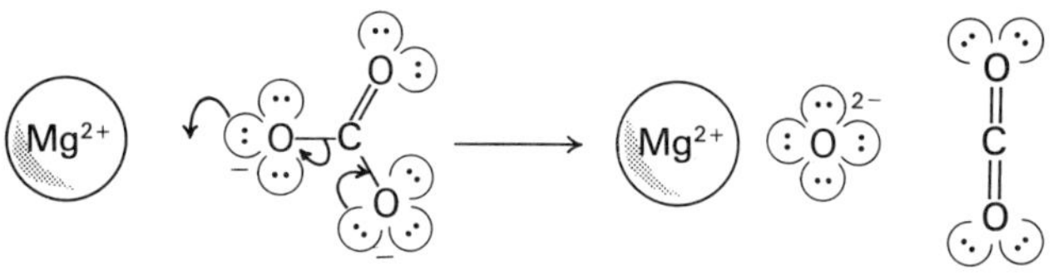

8.3 ALUMINIUM

Structure

Aluminium is in Group III and is less reactive than either sodium or magnesium in Groups I and II. Aluminium atoms have more control of their outer-shell electrons than the atoms of the other two metals. Nonetheless, the degree of control is low enough for the lattice of the element to contain cations in a cloud of delocalized electrons. Compare the ionization energy data given below: an estimate can be made of the ease with which outer-shell electrons are lost from the different atoms.

Na(g)→Na$^+$(g) + e$^-$	Mg(g)→Mg^{2+}(g) + 2e$^-$	Al(g)→Al^{3+}(g) + 3e$^-$
494 kJ mol^{-1}	2186 kJ mol^{-1}	5137 kJ mol^{-1}

Compound formation

Aluminium forms compounds whose degrees of covalency are high. An aluminium cation is small and has a charge of three, so its polarizing power is, therefore, considerable (see the discussion on page 23). Not only does aluminium react with non-metals such as oxygen, chlorine and acid solutions, it also has an extensive organo-metallic chemistry, and forms quite a wide range of complexes. These last two properties are the result of aluminium's tendency to exhibit covalent bonding.

Example 1. Triethylaluminium is used as an ingredient in the Ziegler catalyst for polymerizing ethene and propene (page 193).

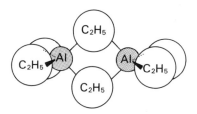

Example 2. Tetrahydridoaluminate ions are formed in ethereal solution when aluminium chloride and lithium hydride are refluxed in ether. These act as a source of hydride ions used to reduce certain classes of organic compound (see page 216).

$$4LiH(s) + AlCl_3(s) \underset{\text{reflux}}{\overset{\text{ether}}{\rightleftharpoons}} LiAlH_4(\text{ether}) + 3LiCl(s)$$

But also, aluminium shows typical metallic reactivity as well, for example,

$$2Al(s) + 3Cl_2(g) \xrightarrow{\text{heat}} Al_2Cl_6(g) \rightleftharpoons 2AlCl_3(s)$$
(see page 24)

Acid-base properties

Amphoteric character. The metal, its oxide and hydroxide are all amphoteric because they react with both acid and alkali. Aluminium metal liberates hydrogen from dilute hydrochloric acid and also from dilute sodium hydroxide, although a persistent oxide layer coating the metal's surface makes the metal appear less reactive than it is. Once the layer is penetrated, reaction is swift:

$$2Al(s) + 6H_3O^+(aq) \longrightarrow 2Al^{3+}(aq) + 6H_2O(l) + 3H_2(g)$$
$$2Al(s) + 2OH^-(aq) + 6H_2O(l) \longrightarrow 2Al(OH)_4^-(aq) + 3H_2(g)$$

The hydroxyl complex $Al(OH)_4^-$ is called the aluminate ion. It is also produced by the reaction of aluminium hydroxide or oxide with alkali.

$$Al_2O_3(s) + 2OH^-(aq) + 3H_2O(l) \xrightarrow{\text{boil}} 2Al(OH)_4^-(aq)$$

Acidic character. The metal salts are acidic as a result of the extensive polarization of the water ligands around the small, highly charged aluminium ion (pH of 1 mol dm^{-3} Al^{3+}(aq) = 3.1).

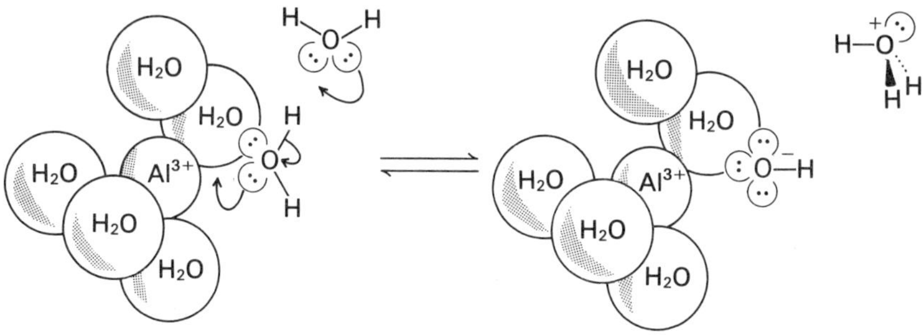

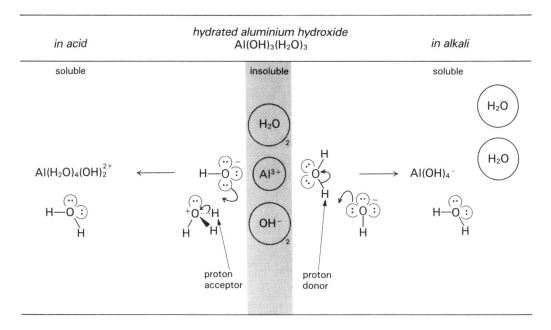

When alkali is added slowly to an aluminium solution, the pH changes as expected for the titration of a weak acid against a strong base (page 92). The pH of the solution remains acidic until the equivalence point is reached. In this case, the product at the equivalence point is hydrated aluminium hydroxide, Al(H$_2$O)$_3$(OH)$_3$(s). The hydrated hydroxide is amphoteric in the Brønsted–Lowry sense: it acts as both proton donor and acceptor.

As base:

$$Al(H_2O)_3(OH)_3 + 3H_3O^+ \longrightarrow Al(H_2O)_6^{3+} + 3H_2O$$

As acid:

$$Al(H_2O)_3(OH)_3 + OH^- \longrightarrow Al(OH)_4^- + 3H_2O$$

Redox properties

Like the Group I and II metals, aluminium metal acts as a reductant. It is not as strong a reductant as sodium or magnesium, and this reflects the gradually increasing control that an atom has over its outer-shell electrons going across a period. In keeping with the increasing control, the three electronegativities also increase: Na = 0.9, Mg = 1.2, Al = 1.5.

$$Al^{3+}/Al \qquad Mg^{2+}/Mg \qquad Na^+/Na$$
$$E^\ominus = -1.66\text{ V} \qquad E^\ominus = -2.38\text{ V} \qquad E^\ominus = -2.71\text{ V}$$

Aluminium is used to reduce iron oxide to iron to make joints. The iron is produced in a molten state and, on cooling, fuses a joint together (the *thermite reaction*).

$$2\overset{0}{Al} + \overset{III}{Fe_2}O_3 \longrightarrow \overset{III}{Al_2}O_3 + 2\overset{0}{Fe}$$

8.4 THE TRANSITION METALS

Position in the Periodic Table

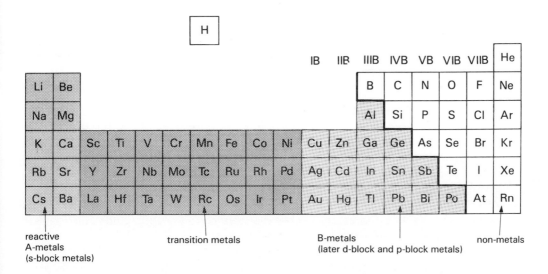

The transition (or d-block) metals get their name from their position on the Periodic Table: they represent the 'transition' from the reactive s-block metals to the less reactive p-block metals. The inner d-subshell structure of a p-block metal atom is complete, whereas that of an s-block metal atom is not. On going across a transition series, it is, therefore, an inner d-subshell that is being filled instead of an outer shell. The reason for this can be explained by the energy diagram shown on page 118 for the orbitals of the first three shells.

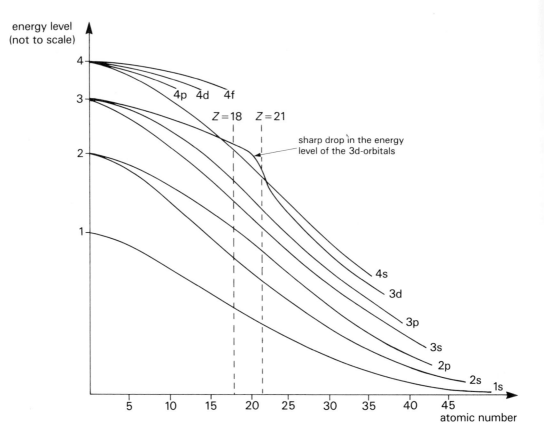

1 As protons are added to the nucleus of an atom, the energy of an orbital decreases because it requires more energy to excite an orbiting electron to the energy level corresponding to its escape from the nucleus.
2 The angular p-orbitals and d-orbitals are less affected than the spherically symmetric s-orbitals by this increasing nuclear charge. As a result, the energy drop of the 4s-orbital becomes greater than that of the inner 3d-orbital at $Z \approx 15$.
3 After the completion of the 3p-orbital at $Z = 18$, the vacant orbital of next lowest energy is the 4s. It is, therefore, filled next according to the aufbau principle (page 9). So potassium has the electronic configuration: $1s^2 2s^2 2p^6 3s^2 3p^6 4s^1$.
4 The energy of the 3d-orbitals drops sharply between $Z = 19$ and $Z = 21$ because they are not shielded at all from the additional protons. The two extra electrons are in an outer orbital.
5 At $Z = 21$, the next available orbital becomes the 3d, whose energy has now dropped below that of the 4s-orbital.

This pattern repeats itself after the filling of the 4p-orbital so that a second transition series occurs during the completion of the 4d-orbitals. Further transition series (including f-block series) also exist. A useful method of remembering the order in which orbitals are filled is shown on page 10.

Compound formation

In effect, all the transition metal atoms have either one or two outer-shell electrons, for example:

chromium $1s^2 2s^2 2p^6 3s^2 3p^6 3d^5 4s^1$
nickel $1s^2 2s^2 2p^6 3s^2 3p^6 3d^8 4s^2$

Chromium is $3d^5 4s^1$ not $3d^4 4s^2$. The Hund principle suggests that $3d^5 4s^1$ is of lower energy, because electron repulsion is minimized. The outer-shell electrons are less well shielded in these atoms than are the outer-shell electrons of a Group I or II atom (because of the poorer screening quality of electrons in the inner 3d-orbitals). There is, therefore, a slight contraction in atomic size going across the transition series, and the electronegativity slowly increases. However, the control of outer-shell electrons is still not good enough to inhibit the metallic character of the elements. The atoms tend to lose electrons to non-metal atoms and form ionic lattices whose degrees of covalency are quite high due to the fairly small sizes of the cations and their charges of two or three.

The partially filled 3d-subshell is largely responsible for two characteristic features shown by transition metals in their compounds:

1 they exhibit variable valency;
2 they form an enormous range of complexes.

Variable valency is the result of low successive ionization energies. The electrons in the 3d-orbitals of a transition metal atom can be lost like those in the 4s; the energy required for the ionization is got back in the energy evolved when new bonds form. For example, compare the first five ionization energies of calcium and vanadium.

atom	ionization energies/kJ mol^{-1}	total/kJ mol^{-1}
calcium	590 1150 4940 6480 8120	21 280
vanadium	648 1370 2870 4600 6280	15 768

As the oxidation state of the transition metal in a compound increases, so also does the degree of covalency of the compound. For example, the bonding in vanadium(V) oxide can be considered initially in terms of the hypothetical existence of V^{5+} ions and

oxidation state	(Sc)	Ti	V	Cr	Mn	Fe	Co	Ni	(Cu)
II		•	•	•	•	•	•	•	•
III	•	•	•	•	•	•	•	•	
IV		•	•	•	•	•	•	•	
V			•	•	•	•			
VI				•	•	•			
VII					•				

Cu(II) is 'transitional' because it contains an incomplete d-subshell $3d^9$

O^{2-} ions. However, the polarizing power of the tiny, highly charged V^{5+} ions is enormous and, therefore, the degree of covalency is high. The most common oxidation states of the transition metals in their compounds are shown in the table on page 119.

Complex formation. A complex contains a transition metal atom (usually in a positive oxidation state) to which a number of ligands are bonded. The ligands are particles each with a lone pair of electrons co-ordinated to the central atom. Sometimes, the ligands are basic molecules (e.g. H_2O, NH_3), in which case, the complex itself is cationic; however, most ligands are negatively charged ions (e.g. Cl^-, CN^-, O^{2-}), and, therefore, the complex is often anionic.

Different ligands stabilize different ranges of transition metal oxidation states. In the lower oxidation states (−I, 0, I), the central transition metal atom still has considerable electron density in the 3d-orbitals. A ligand must, therefore, be able to accept these electrons in order to stabilize the oxidation state (e.g.: CO). Conversely, the highest oxidation state is reached when the transition metal loses all its 3d- and 4s-electrons. Ligands that stabilize these states must be both σ- and π-donors (e.g. O^{2-}). The general pattern is summarized by the table below.

low oxidation states	medium oxidation states	high oxidation states
$M \underset{\pi}{\overset{\sigma}{\rightleftarrows}} L$	$M \overset{\sigma}{\leftarrow} L$	$M \underset{\pi}{\overset{\sigma}{\rightleftarrows}} L$
e.g. $\overset{0}{Cr(CO)_6}$	e.g. $\overset{III}{Cr(NH_3)_6^{3+}}$	e.g. $\overset{VI}{CrO_4^{2-}}$

The bonding between a transition metal atom and the ligands surrounding it cannot be fully explained by the covalent model. Part of the force of attraction is simply an electrostatic one in which no electron transfer occurs at all. A combination of the simple electrostatic model (crystal field theory) and the covalent model (molecular orbital theory) is called the ligand field theory.

Further effects of an incomplete d-subshell

There are three other properties of transition metal compounds that arise from the effects of the incomplete d-subshells:

1 the compounds are usually coloured;
2 they sometimes exhibit paramagnetism;
3 they often act as catalysts.

Colour. The electrostatic crystal field set up by the ligands around a transition metal atom changes the energy-level of the five 3d-orbitals. For example, consider an octahedral field:

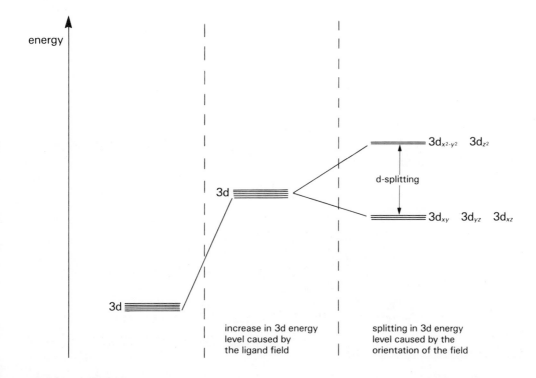

octahedral arrangement of ligands around a central metal cation

The arrangement of the ligands raises the energy of the $3d_{x^2-y^2}$ and $3d_{z^2}$ by a greater amount than that of the others, because these two orbitals point out along the axes, directly at the negative charge clouds of the ligands. Thus, the 3d-orbitals energy is split: the term *d-splitting* is used to describe the effect.

The magnitude of the d-splitting is in the energy band corresponding to a quantum of visible light ($E = h\nu_{visible}$; see page 6). Certain frequencies are, therefore, absorbed from white light (sunlight) as transitions take place from the lower 3d level to the higher. The light that is transmitted is, therefore, coloured.

Magnetism is associated with substances whose atoms possess unpaired electrons. An electron has spin momentum within its orbital and, when the orbital is full, the spins of the two electrons are opposed.

However, if the electron is unpaired, a small magnetic field is set up by the spinning charge. When all these small fields act together, the substance exhibits paramagnetic properties, aligning itself in the direction of an applied external magnetic field. Transition metals and their compounds often have unpaired electrons in the 3d-orbitals, and these give rise to paramagnetism. The effect is so marked in iron that the term ferromagnetic is used. For example, in iron(III) chloride, iron is present as $3d^5$ ions.

$$[Ar]3d^6 4s^2 \longrightarrow [Ar]3d^5 + 3e^-$$

To minimize electronic repulsion, the five electrons are likely to be assigned one each to the five different d-orbitals, giving five unpaired spins.

As catalysts. The presence of partially full (and empty) 3d-orbitals provides reactant particles with the means of forming weak bonds to the catalyst. This brings the reactant particles closer together and also reduces the amount of energy required to break the bonds in the particles. Examples include the use of vanadium(V) oxide in the oxidation of sulphur dioxide and of iron in the synthesis of ammonia.

8.5 CHROMIUM AND IRON

Structure

At room temperature, chromium and iron both have body-centred cubic lattices of close packed cations in a cloud of delocalized electrons. There is delocalization of both the 4s and some of the 3d electrons as well. The electronic configurations, successive ionization energies and electronegativities of the two metals are shown below. These illustrate the general patterns described earlier.

metal	electronic configuration	ionization energies / kJ mol^{-1}	electro-negativity
chromium	$1s^2 2s^2 2p^6 3s^2 3p^6 4s^1 3d^5$	653 1590 2990 4770 7070 8700	1.6
iron	$1s^2 2s^2 2p^6 3s^2 3p^6 4s^2 3d^6$	762 1560 2960 5400 7620 10 100	1.8
compare with			
calcium	$1s^2 2s^2 2p^6 3s^2 3p^6 4s^2$	590 1150 4940 6480 8120 10 700	1.0

Compound formation

Chromium and iron show all the properties typical of a transition metal: they form a wide range of coloured compounds in a number of different oxidation states and they show a marked tendency to exist as complex rather than simple salts.

Lower oxidation states are stabilized only by π-donor ligands, such as carbon monoxide molecules and cyanide ions. The zero oxidation state occurs in the metal 'carbonyls' of which $Cr(CO)_6$ and $Fe(CO)_5$ are the simplest (see page 120).

Medium oxidation states are reached when the metals combine with non-metals or with acids. The two most stable states are II and III, although chromium(II) is very easily oxidized to chromium(III). It is, however, possible to prepare a red, hydrated ethanoate complex of chromium(II) that is fairly resistant to air oxidation. The equations given below provide a summary of the formation of medium oxidation state compounds.

Chromium(II) or iron(II)

$$M(s) + 2HCl(g) \longrightarrow MCl_2(s) + H_2(g)$$

$$M(s) + 2H_3O^+(aq) \longrightarrow M^{2+}(aq) + 2H_2O(l) + H_2(g)$$

but $Cr^{2+}(aq)$ is unstable unless precipitated via

$$2Cr^{2+}(aq) + 4CH_3COO^-(aq) + 2H_2O(l) \longrightarrow Cr_2(CH_3COO)_4(H_2O)_2(s)$$

The structure of the chromium(II) complex appears to indicate the likelihood of metal-metal bonding.

Chromium(III) or iron(III)

$$2M(s) + 3Cl_2(g) \xrightarrow{heat} 2MCl_3(s)$$

$$4M(s) + 3O_2(g) \xrightarrow{heat} 2M_2O_3(s)$$

Higher oxidation states are reached when solutions of the metal compounds are treated with a strong oxidant in concentrated alkali. Chromium is much more readily oxidized than iron, as might be expected from the ionization energy values shown on page 122. In both cases, the highest stable oxidation state is six. For chromium, it is most readily attained by reacting chromium(III) chloride with hydrogen peroxide in concentrated alkali. An intermediate peroxide complex (CrO_5) can be extracted into an ether layer if the reaction mixture is shaken with ether. CrO_5 in ether is a pale blue colour.

$$2Cr^{3+}(aq) + 6OH^-(aq) + 7H_2O_2(aq) \xrightarrow{warm} 2[CrO_5(aq)] + 10H_2O(l)$$

$$\updownarrow ether$$

$$2CrO_5(ether) \text{ blue}$$

The peroxide complex is easily hydrolysed to chromate(VI) ions.

$$CrO_5(aq) + 2OH^-(aq) + H_2O(l) \longrightarrow CrO_4^{2-}(aq) + 2H_2O_2(aq)$$

Under acid conditions, chromate(VI) ions condense to give dichromate(VI) ions.

yellow chromate (VI) $+ 2H_3O^+ \rightleftharpoons$ orange dichromate (VI) $+ 3H_2O$

For iron, the VI state can be prepared by the anodic oxidation of a soft iron electrode in an electrolysis cell containing concentrated alkali. A high current density is passed and the following anode reaction takes place.

$$Fe(s) + 8OH^-(aq) \longrightarrow FeO_4^{2-}(aq) + 4H_2O(l) + 6e^-$$
purple

Acid-base properties

The metals and their simple oxides and hydroxides show the basic properties typical of any metal. The oxide and hydroxide of chromium(III) are amphoteric (like those of aluminium(III), page 116), for example:

$$\underset{metal}{M(s)} + \underset{acid}{2H_3O^+(aq)} \longrightarrow \underset{salt}{M^{2+}(aq)} + 2H_2O(l) + \underset{hydrogen}{H_2(g)}$$

$$\underset{base}{Fe(OH)_2(s)} + \underset{acid}{2H_3O^+(aq)} \longrightarrow \underset{salt}{Fe^{2+}(aq)} + \underset{water}{4H_2O(l)}$$

$$\underset{acid}{Cr(OH)_3(H_2O)_3} + \underset{base}{3OH^-(aq)} \longrightarrow \underset{salt}{Cr(OH)_6^{3-}(aq)} + \underset{water}{3H_2O(l)}$$

Compounds of the metals in higher oxidation states are all acidic by hydrolysis. The two most typical examples are chromium(VI) oxide and dichlorochromium(VI) oxide: the chromates and dichromates are the salts of the acid produced by the hydrolysis of these compounds. The reverse of the hydrolysis reaction is used to prepare a sample of chromium(VI) oxide

$$Cr_2O_7^{2-}(aq) + 3H_2SO_4(l) \longrightarrow 2CrO_3(s) + 3HSO_4^-(aq) + H_3O^+(aq)$$

A concentrated solution of potassium dichromate is treated with the dehydrating agent, concentrated sulphuric acid. Chromium(VI) oxide is precipitated as a red solid which under aqueous conditions is rapidly hydrolysed. It has some similarity to sulphur(VI) oxide.

$$2CrO_3(s) + 3H_2O(l) \longrightarrow Cr_2O_7^{2-}(aq) + 2H_3O^+(aq)$$

Redox properties

A transition metal forms a wider range of oxidation states (and therefore has a more extensive redox chemistry) than a non-transition metal. However the greater control that a typical transition metal atom has over its outer-shell electrons, makes it a weaker reductant than a typical s-block metal. Compare the redox potentials below.

$$Ca^{2+}(aq) + 2e^- \rightleftharpoons Ca(s) \quad E^\ominus = -2.87 \text{ V}$$
$$Cr^{2+}(aq) + 2e^- \rightleftharpoons Cr(s) \quad E^\ominus = -0.91 \text{ V}$$
$$Fe^{2+}(aq) + 2e^- \rightleftharpoons Fe(s) \quad E^\ominus = -0.44 \text{ V}$$

Both iron(II) and chromium(II) are useful reductants in aqueous solution, whereas calcium(II) has no reducing properties because it cannot be converted to calcium(III). Chromium(II) is, in fact, one of the strongest aqueous reducing reagents, whereas iron(II) is sufficiently weak to allow iron(III) some oxidizing properties. For example, chromium(II) is readily oxidized by iron(III).

$$Fe^{3+}(aq) + e^- \rightleftharpoons Fe^{2+}(aq) \quad E^\ominus = +0.77 \text{ V}$$
$$Cr^{3+}(aq) + e^- \rightleftharpoons Cr^{2+}(aq) \quad E^\ominus = -0.41 \text{ V}$$
$$\therefore Cr^{2+}(aq) + Fe^{3+}(aq) \longrightarrow Cr^{3+}(aq) + Fe^{2+}(aq)$$

In the higher oxidation state VI, chromium and iron are both strongly oxidizing and, for example, liberate iodine from an acidified iodide solution.

$$I_2(aq) + 2e^- \rightleftharpoons 2I^-(aq) \quad \| \quad Cr_2O_7^{2-}(aq) + 14H_3O^+(aq) + 6e^- \rightleftharpoons 2Cr^{3+}(aq) + 21H_2O(l)$$

$$E^\ominus = +0.54 \text{ V} \qquad\qquad E^\ominus = +1.33 \text{ V}$$

electron flow

Therefore iodide ions give up electrons while dichromate ions accept them.

8.6 THE B-METALS

Position in the Periodic Table

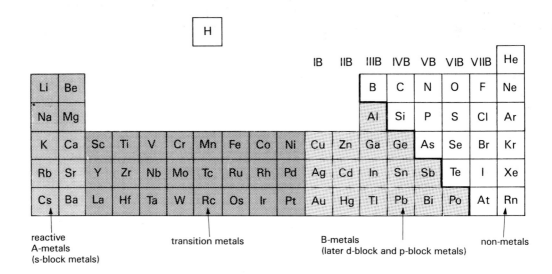

The B-metals occur at the foot of the p-block and in the last two groups of the d-block. In a B-metal atom, the inner d-subshell structure (where it exists) is full, whereas it is empty, or partially empty, in an A-metal atom. For example, the electronic configurations of potassium(IA) and copper(IB) are $1s^2 2s^2 2p^6 3s^2 3p^6 4s^1$ and $1s^2 2s^2 2p^6 3s^2 3p^6 3d^{10} 4s^1$. The only difference between the two is the full set of d-orbitals in the copper atom; both have one outer-shell electron in a 4s-orbital and hence, are Group I metals. B-metals tend to be less reactive than A-metals, because their atoms have more control of their outer-shell electrons. The full inner d-subshell does not act as a very effective shield for the outer-shell from the increased nuclear charge. For example, compare the data below for potassium and copper.

metal	electronic configuration	ionization energy/$kJ\ mol^{-1}$	electro-negativity	$E^{\ominus}$/volt
potassium	$1s^2 2s^2 2p^6 3s^2 3p^6 4s^1$	418	0.8	−2.92
copper	$1s^2 2s^2 2p^6 3s^2 3p^6 3d^{10} 4s^1$	745	1.9	+0.34

The emergence of B-metallic character at the foot of the p-block is largely due to the *inert pair effect*. This is the name given to the withdrawal of the outer s-electrons into the core of an atom. The effect is caused by a decrease in the efficiency of the screening of an atom's outer shell as the number of electrons in the angular d-orbitals and f-orbitals increases with increasing atomic number. The spherically symmetric outer s-orbital is more affected by the poor screening than are the outer p-orbitals, and this has two results.

1 The outer s-electron pair behaves more like an inner-shell pair, because it is withdrawn into the atomic core. Hence, it is called an *inert pair*.
2 The inert pair shields the outer p-electrons from the nuclear charge, even though it is part of the same shell. This decreases the control of the nucleus over the outer p-electrons which are, therefore, more readily lost. For example, the first two ionization energies of the B-metal, tin, are not much greater than those of the reactive A-metal from the same period, strontium.

metal	first IE/kJ mol^{-1}	second IE/kJ mol^{-1}
tin	707	1410
strontium	548	1060

A comparison of A-metal and B-metal chemistry

The less reactive nature of the B-metals is well illustrated by comparing the chemistries of potassium and copper ($4s^1$) or calcium and zinc ($4s^2$). For example, copper does not displace hydrogen, and resists air oxidation, whereas the following reactions of potassium occur violently.

$$2K(s) + 2H_2O(l) \longrightarrow 2K^+(aq) + 2OH^-(aq) + H_2(g)$$
$$2K(s) + O_2(g) \longrightarrow K_2O_2(s)$$

Copper, in fact, has transition metal properties because its atoms tend to lose *two* electrons in bonding; Cu^{2+} ions have the configuration $3d^9$. Copper(II) compounds, therefore, show the characteristics typical of transition metal compounds.

Copper(I) is unstable in aqueous solution, unless complexed by π-acceptor ligands (such as cyanide ions, see page 120), or by iodide ions. It disproportionates to give copper(II) and metallic copper. (Disproportionation is a redox change in which a substance is simultaneously both oxidized and reduced.) The redox potentials of copper indicate the likelihood of the disproportionation.

$$Cu^{2+}(aq) + e^- \rightleftharpoons Cu^+(aq) \qquad\qquad Cu^+(aq) + e^- \rightleftharpoons Cu(s)$$

$$E^\ominus = +0.15\text{ V} \qquad\qquad E^\ominus = +0.52\text{ V}$$

electron flow

$$Cu^+(aq) + Cu^+(aq) \longrightarrow Cu^{2+}(aq) + Cu(s)$$

Copper(I) iodide is insoluble and can be produced from a copper(II) solution by reduction using iodide ions.

$$2\overset{II}{Cu}{}^{2+}(aq) + 4\overset{-I}{I}{}^-(aq) \longrightarrow 2\overset{I}{Cu}I(s) + \overset{0}{I}_2(aq)$$

Not only is a B-metal less reactive than the A-metal of the same period, there are also significant differences in the compounds formed. The B-metals often combine in more than one oxidation state; the degree of covalency of their compounds is higher, and they show a marked tendency towards complex formation.

Variable oxidation state. A B-metal does not exhibit the range of states typical of a transition metal. There are only two stable states: N and $(N - 2)$ where N is the group number. The first state is reached when all the electrons are used for bonding, while the second is reached when the inert pair effect is strongly in evidence. The lower oxidation state becomes more stable going down a group, because the inert pair effect is more marked in the lower group members. For example, indium(I) and tin(II) are quite strongly reducing, whereas thallium(III) and lead(IV) are useful oxidants. This reflects the tendency of indium and tin to exist in the higher oxidation state, but for thallium and lead to exist in the lower. For example,

tin(II) as reductant:

$$C_6H_5NO_2 + 3SnCl_2^{II} + 7HCl \longrightarrow C_6H_5NH_3^+Cl^- + 3SnCl_4^{IV} + 2H_2O$$

lead(IV) as oxidant:

$$PbO_2^{IV} + 4HCl^{-I} \longrightarrow PbCl_2^{II} + 2H_2O + Cl_2^{0}$$

The degree of covalency in a B-metal compound is high because of the large polarizing power of the cations present. The poor screening of the B-metal nuclei is chiefly responsible for larger polarizing powers than might be expected from an inspection of cation sizes. For example, the lattice structure of magnesium (Group IIA) oxide is a 6:6 ionic arrangement like that of sodium chloride. However, zinc (Group IIB) oxide has considerably more covalent character and exists in a 4:4 diamond-type lattice, even though the radii of a zinc ion and a magnesium ion are almost equal (0.75×10^{-10} and 0.65×10^{-10} m).

Similarly, the observed lattice energy of silver chloride is greater than that of sodium chloride, despite the fact that the silver ions are larger than the sodium ions. By the inverse square law, it might be expected that sodium chloride's lattice energy would be larger. However, the degree of covalency of silver chloride is greater, and the extra bond energy resulting from this leads to a higher value for the lattice energy.

Complex formation is favoured for similar reasons: the tendency of B-metal atoms to show appreciable covalent character in their bonding is well illustrated by the range of B-metal complexes. Some typical examples are given below.

1 Silver chloride dissolves in aqueous ammonia, but not in water.

$$AgCl(s) + 2NH_3(aq) \longrightarrow Ag(NH_3)_2^+(aq) + Cl^-(aq)$$

2 Lead(II) chloride is soluble in concentrated hydrochloric acid, but not in cold water.

$$PbCl_2(s) + 2Cl^-(aq) \longrightarrow PbCl_4^{2-}(aq)$$

3 Many B-metal hydroxides are amphoteric (see page 85).

$Al(OH)_3(s) + OH^-(aq) \longrightarrow Al(OH)_4^-(aq)$

$Sn(OH)_2(s) + 2OH^-(aq) \longrightarrow Sn(OH)_4^{2-}(aq)$

$Pb(OH)_2(s) + 2OH^-(aq) \longrightarrow Pb(OH)_4^{2-}(aq)$

$Zn(OH)_2(s) + 2OH^-(aq) \longrightarrow Zn(OH)_4^{2-}(aq)$

4 Copper hydroxide dissolves in aqueous ammonia.

$\underset{\text{pale blue}}{Cu(OH)_2(s)} + 4NH_3(aq) \longrightarrow \underset{\text{dark blue}}{Cu(NH_3)_4^{2+}(aq)} + 2OH^-(aq)$

9 Chemistry of the Non-metals

9.1 GROUP VII: THE HALOGENS

Structure

A halogen atom has a core charge of seven and has seven outer-shell electrons held tightly to the core. The first ionization energies of the halogens are close to the peaks on the graph shown on page 109; only the noble gases have higher ionization energies.

The halogens are diatomic non-metals at room temperature. The molecules are held together by fairly weak σ-bonding: the bond strength decreases going down the group due to the increased distance of the nuclei from the bonding electrons in each case. The bond strength in F—F is an exception: the repulsion between the lone pairs of each atom is greatest for a fluorine molecule.

molecule	bond length/nm	bond strength/kJ mol^{-1}
F—F	0.142	158
Cl—Cl	0.199	242
Br—Br	0.228	193
I—I	0.267	151

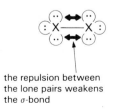

the repulsion between the lone pairs weakens the σ-bond

Compound formation

The high electron affinities and electronegativities of the halogens suggest that they are likely to be found combined in a negative oxidation state. They are the most reactive group of non-metals and combine with nearly every other element. With the exception of the oxycompounds and the interhalogen compounds, the halogens are always found combined in an oxidation state of −I.

With metals

$$2Na(s) + \overset{0}{Cl_2}(g) \longrightarrow Na\overset{-I}{Cl}(s)$$

$$2Al(s) + 3\overset{0}{I_2}(s) \xrightarrow{heat} 2Al\overset{-I}{I_3}(s)$$

Metal halides are predominantly ionic and contain halide ions, X^-. The degree of covalency increases as the atomic number increases because of the increase in the anion size (Fajans' rules, page 16). Similarly, the degree of covalency increases as the charge of the cation increases and the size of the cation decreases.

With non-metals

$$S_8(s) + 4\overset{0}{Cl_2}(g) \longrightarrow 4S_2\overset{-I}{Cl_2}(l)$$

$$P_4(s) + 3\overset{0}{Br_2}(l) \longrightarrow 4P\overset{-I}{Br_3}(s)$$

Like metal halides, non-metal halides are also readily formed by direct combination. They are molecular compounds and there is charge separation along the σ-bonds. Phosphorus(III) bromide and hydrogen chloride are illustrated below.

Oxycompounds and interhalogens contain a halogen in a positive oxidation state and they are formed by all the halogens except fluorine. It is not possible for fluorine to have a positive oxidation state chemistry, because it is the most electronegative of all the elements. Fluorine does form two oxides (F_2O and F_2O_2) and a range of interhalogens but, in all these compounds, the element has an oxidation state of −I. Chlorine, bromine and iodine are able to expand their bonding octet by using the vacant d-orbitals of the outer shell. The energy required for the promotion of electrons into these higher-energy, vacant orbitals is offset by a larger amount of energy released when further bonds are formed. Halogens combine in positive oxidation states I, III, V and VII. Two examples are shown overleaf.

Example 1 BrF₅

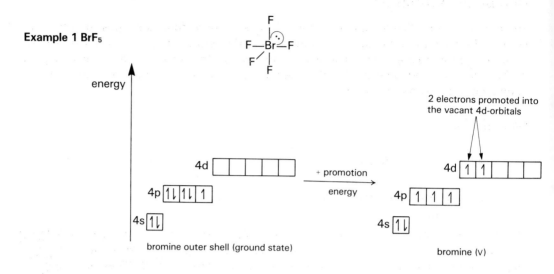

Example 2 NaClO₄

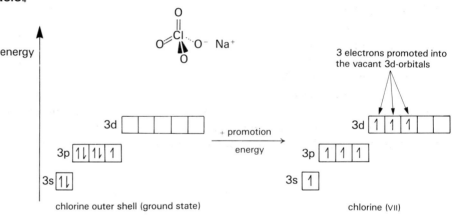

An interhalogen is prepared by direct combination like any other non-metal halide. The oxycompounds are prepared by the action of alkali on a halogen.

$$3\overset{0}{Cl}_{2(g)} + 6OH^-_{(aq)} \longrightarrow 3\overset{-I}{ClO}^-_{(aq)} + 3\overset{-I}{Cl}^-_{(aq)} + 3H_2O_{(l)}$$

and then,

$$3\overset{I}{ClO}^-_{(aq)} \longrightarrow \overset{V}{ClO}_3^-_{(aq)} + 2\overset{-I}{Cl}^-_{(aq)}$$

The reactions shown above are all examples of disproportionation because, in each case, a substance is undergoing simultaneous oxidation and reduction. Solid sodium chlorate(V) disproportionates further on heating to give chlorate(VII).

$$4Na\overset{V}{Cl}O_3(s) \xrightarrow{heat} Na\overset{-I}{Cl} + 3Na\overset{VII}{Cl}O_4 \xrightarrow[heat]{strong} 4Na\overset{-I}{Cl} + 6\overset{0}{O_2}(g)$$

Acid-base properties

Metal halides are usually soluble in water and give solutions whose pH depends on the extent of the hydrolysis of the cations present (see page 116). Fluorides, however, are slightly alkaline due to the weak basic strength of the fluoride ion.

$$F^-(aq) + H_2O(l) \rightleftharpoons HF(aq) + OH^-(aq) \quad K_b = 10^{-11.75} \text{ mol dm}^{-3}$$

Hydrogen fluoride is an anomalously weak acid compared with the other hydrogen halides, and so its conjugate base has significant basicity. The weakness of hydrogen fluoride as an acid is due to two factors:

1. the H—F bond broken is stronger than the O—H bond formed (463 kJ mol^{-1});
2. molecular H—F is stabilized by hydrogen bonding to water molecules (see page 20).

halogen (X)	H—X bond strength/kJ mol^{-1}	K_a/mol dm^{-3}
fluorine	562	1.8×10^{-4}
chlorine	431	$\sim 10^7$
bromine	366	$\sim 10^8$

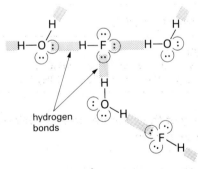

The tendency for hydrogen fluoride to exhibit hydrogen bonding is so marked that a stable salt of formula NaHF$_2$ can be prepared. The HF$_2^-$ ion contains a fluoride ion hydrogen bonded to a hydrogen fluoride molecule. The formation of this ion in aqueous solutions of hydrogen fluoride further affects the acidic strength.

Non-metal halides are almost invariably acidic in water. Hydrolysis occurs to produce oxyacids and halide ions. For example,

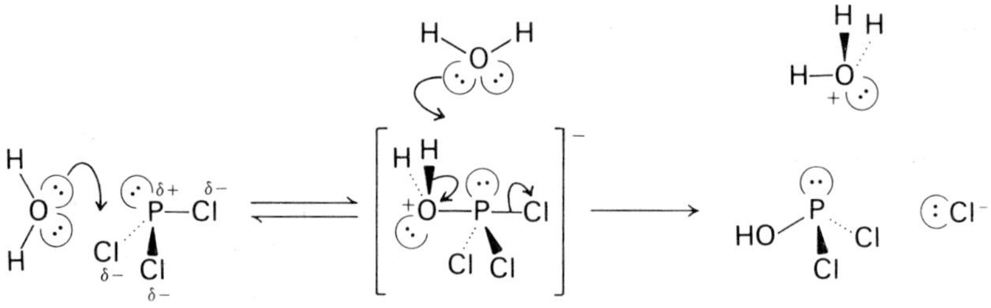

Further hydrolysis takes place, so that the overall reaction is as follows:

$$PCl_3(s) + 6H_2O(l) \longrightarrow H_3PO_3(aq) + 3Cl^-(aq) + 3H_3O^+(aq)$$

Halogen oxides and interhalogens are hydrolysed in the same way. The products are halogen oxyacids. For example,

$$ClF_3(l) + 3H_2O(l) \longrightarrow ClO_2^-(aq) + 3HF(aq) + H_3O^+(aq)$$
$$I_2O_5(s) + 3H_2O(l) \longrightarrow 2IO_3^-(aq) + 2H_3O^+(aq)$$

Redox properties

The elements themselves are quite strong oxidants and tend to react to produce compounds containing the halogen in the −I state. The positive oxidation state compounds are also oxidants for the same reason. The redox potentials for chlorine and the chlorates (I) and (V) are given below.

redox couple	$E^\ominus/V$
$Cl_2(g) + 2e^- \rightleftharpoons 2Cl^-(aq)$	+1.36
$2HOCl(aq) + 2H_3O^+(aq) + 2e^- \rightleftharpoons Cl_2(g) + 4H_2O(l)$	+1.64
$2ClO_3^-(aq) + 12H_3O^+(aq) + 10e^- \rightleftharpoons Cl_2(g) + 18H_2O(l)$	+1.47

A solution of chlorine (or any of the chlorates in acid) oxidizes, for example, iron(II) to iron(III) or hydrogen peroxide to oxygen.

$$3H_2\overset{-I}{O}_2(aq) + \overset{V}{Cl}O_3^-(aq) \longrightarrow 3H_2O(l) + \overset{-I}{Cl}^-(aq) + 3\overset{0}{O}_2(g)$$

Fluorine is a much stronger oxidant than any of the other halogen couples. The redox potential is +2.87 volts and is twice that of chlorine. This large difference can be explained by considering the energy factors involved.

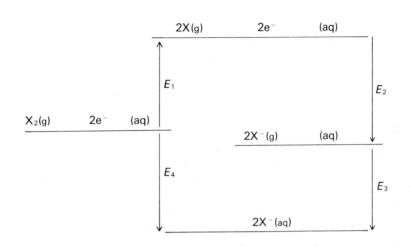

E_1 is the bond dissociation energy, which is anomalously small for F—F (see table on page 130).

E_2 is twice the electron affinity of the halogen. There is little variation in the electron affinity of the different halogens; the values range from 315 to 360 kJ mol^{-1}.

E_3 is twice the hydration energy of the halide ion. This is much larger for fluorine because the charge density of the ion is higher as a result of its small size.

E_4 is proportional to the redox potential and is given by $(E_1 + E_2 + E_3)$. The larger the value of E_4, the stronger is the oxidizing power of the halogen. From the above arguments, it can be seen that two of the three factors significantly favour fluorine.

Complexing properties

Halide ions act as good σ-donor ligands (see page 120). They can stabilize a wide range of different transition metal, B-metal and even non-metal atoms in positive oxidation states. Some examples are shown below.

transition metals	B-metals	non-metals
TiF_6^{2-}	AlF_6^{3-}	SiF_6^{2-}
$MnCl_4^-$	$PbCl_4^{2-}$	ICl_4^-

9.2 GROUP VI: OXYGEN AND SULPHUR

Structure

Oxygen and sulphur atoms have six outer-shell electrons and these are strongly attracted to the nucleus like those of the halogens. Their electronic configurations are shown below.

element	electronic configuration	electronegativity
oxygen	$1s^2 2s^2 2p^4$	3.5 (compare F = 4.0)
sulphur	$1s^2 2s^2 2p^6 3s^2 3p^4$	2.5 (compare Cl = 3.0)

There are two points to note about these configurations

1. The two vacancies that an oxygen atom has are in the 2p-orbitals. This leads to the ready formation of a double bond (a σ-bond and a π-bond) because of the extensive π-overlap that an oxygen 2p-orbital can undergo with either the 2p or 3d-orbital of another atom.
2. Although sulphur also has two vacancies in its outer p-subshell, pπ-bonding rarely occurs. Instead, electrons are often promoted into the empty d-subshell of a sulphur atom's outer shell. The promotion energy is got back in the form of the extra bond energy released (see page 139).

The reluctance of sulphur atoms to undergo 3pπ-overlap is illustrated immediately by the molecular structure of the two elements. Oxygen exists as 2pπ-stabilized molecules, whereas sulphur exists as σ-bonded S$_8$ rings.

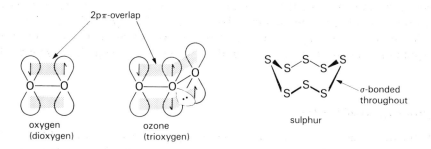

Both elements exhibit allotropy; the enantiotropy of sulphur is discussed on page 51. Liquid sulphur is an unusual substance which thickens on heating. As their kinetic energy increases, the S$_8$ rings start to break up and their fragments join together to form long chains. These molecular chains become longer and more tangled as the temperature rises. Above a certain temperature, the chains themselves begin to break up into smaller units and the liquid then becomes more mobile again. This is an example of dynamic allotropy. If the thick, red liquid sulphur is shock-cooled, a lattice of tangled macromolecular chains forms (called plastic sulphur).

Unlike sulphur, the allotropy of oxygen is monotropic: ozone is the less stable form under all conditions. It is formed as a result of a homolytic reaction induced in oxygen either by intense ultraviolet radiation (as in the upper atmosphere), or by electrical discharge through the gas.

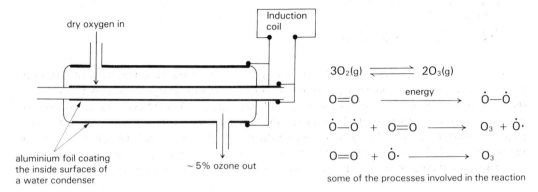

some of the processes involved in the reaction

Compound formation

Oxygen and sulphur are quite strongly electronegative elements and tend to combine with metals and less reactive non-metals to produce compounds containing the Group VI element in a negative oxidation state. Only fluorine is more electronegative than oxygen and so the chemistry of oxygen is dominated by negative oxidation states. There are four characteristic bonding patterns shown by combined oxygen atoms:

electron pairs			structure
lone pair	σ-pair	π-pair	
4	0	0	oxide ion
3	1	0	
2	2	0	
2	1	1	

Sulphur, on the other hand, can expand its octet (like chlorine and bromine, page 132), and have a positive oxidation state chemistry when combined with oxygen or the more reactive halogens. In particular, a wide range of oxyanions exists.

Negative oxidation states. The most important and common state is −II. Metal oxides and sulphides are formed by direct combination between a metal and oxygen or sulphur. these contain O^{2-} or S^{2-} ions.

$$8Zn(s) + \overset{0}{S}_8(s) \longrightarrow 8Zn\overset{-II}{S}(s)$$

$$2Al(s) + 3\overset{0}{O}_2(g) \longrightarrow 2Al_2\overset{-II}{O}_3(s)$$

The degree of covalency of these compounds is considerable because of the high polarizability of the anions. Oxygen also forms peroxides and superoxides with the more reactive metals such as sodium and potassium (page 111). These are examples of catenated compounds, so called because they contain atoms of the same element bonded together into rings or chains and onto which other atoms can be attached (Latin *catena* = chain). There are also examples of catenated sulphur compounds such as the thio-compounds (page 142) or the polysulphides produced when sulphur is boiled in alkali.

$$2S_8 + 8OH^- \xrightarrow{\text{boil}} 3[S-S-S-S-S]^- + ^-O-\overset{O}{\underset{O}{S}}-O + 4H_2O$$

a typical polysulphide chain; other chain-lengths are also produced as well

With non-metals, sulphur is occasionally combined in a negative oxidation state, but usually the combined non-metal is either oxygen or a halogen so that sulphur has a positive oxidation state. Oxygen, on the other hand, forms an enormous range of non-metal oxides containing oxygen in the −II state. These oxides and their corresponding oxyacids and oxyanions are discussed separately for each non-metal. A few of the different types of non-metal oxides and oxyanions are shown below for comparison.

two σ-bonds two lone pairs	one σ-bond one π-bond	one σ-bond three lone pairs

Most oxides can be prepared by direct combination, and most oxyacids and oxyanions by the hydrolysis of these oxides. It is also possible to prepare peroxides (see page 111) containing oxygen nominally in oxidation state −I.

$$Ba + O_2 \xrightarrow{\text{burn in excess } O_2} Ba^{2+}\ [\colon\!\!\overset{..}{\underset{..}{O}}\!-\!\overset{..}{\underset{..}{O}}\!\colon]^{2-} \xrightarrow{\text{dil. } H_2SO_4} BaSO_4(s) + H\!-\!\overset{..}{\underset{..}{O}}\!-\!\overset{..}{\underset{..}{O}}\!-\!H$$

Positive oxidation states are not possible for oxygen but a sulphur atom can promote either one or two electrons to the vacant 3d-orbitals, and thus reach oxidation states of +IV or +VI. The promotion energy is offset by the extra bond energy evolved.

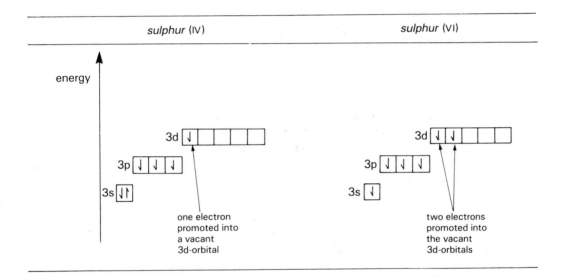

For example, sulphur combines with oxygen to give sulphur(VI) and sulphur(IV) oxide, better known as sulphur trioxide and sulphur dioxide. Similar reactions occur with fluorine, and also oxyhalides can be produced. The structures of three of these molecules are shown below.

$$2\overset{IV}{SO_2}(g) + O_2(g) \underset{}{\overset{V_2O_5}{\rightleftharpoons}} 2\overset{VI}{SO_3}(g)$$

$$PCl_5(s) + \overset{IV}{SO_2}(g) \xrightarrow{\text{heat}} POCl_3(g) + \overset{IV}{SOCl_2}(g)$$

Acid-base properties

Metal oxides and sulphides are bases. Oxide and sulphide ions are proton acceptors and can even accept protons from water molecules if the metal cations in the lattice have a low enough polarizing power (Group I).

$$Na_2O(s) + H_2O(l) \longrightarrow 2Na^+(aq) + 2OH^-(aq)$$

$$Na_2S(s) + H_2O(l) \rightleftharpoons 2Na^+(aq) + OH^-(aq) + HS^-(aq)$$

Most metal oxides and sulphides neutralise acidic solutions, dissolving in them to produce salts. For example,

$$ZnO(s) + 2H_3O^+(aq) \longrightarrow Zn^{2+}(aq) + 3H_2O(l)$$

$$FeS(s) + 2H_3O^+(aq) \longrightarrow Fe^{2+}(aq) + 2H_2O(l) + H_2S(g)$$

The last reaction is the standard method for preparing hydrogen sulphide gas in the laboratory.

Non-metal oxides are acidic. The acid-base properties of metal and non-metal oxides are best contrasted by considering the approach of a water molecule to an oxide lattice. (Initially, the ionic model is used to describe the bonding in the lattice). A water molecule interacts with an oxide lattice in one of two ways, depending on the degree of polarization of the oxide ions. The two ways are illustrated below for the extreme cases of zero and extensive polarization.

zero polarization: a lattice of cations and oxide anions

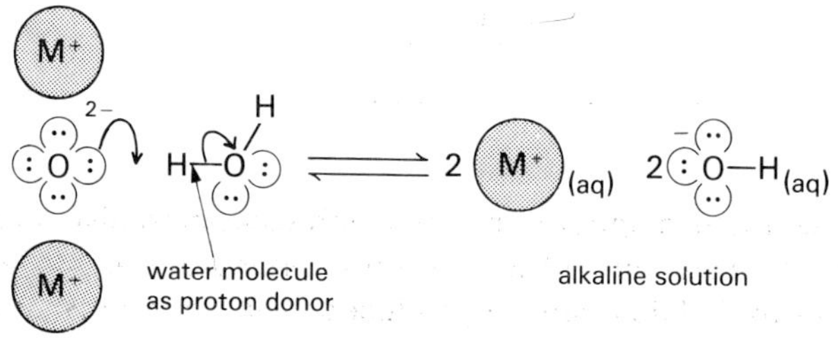

water molecule as proton donor

alkaline solution

extensive polarization: a lattice of molecules containing covalently-bonded oxygen atoms

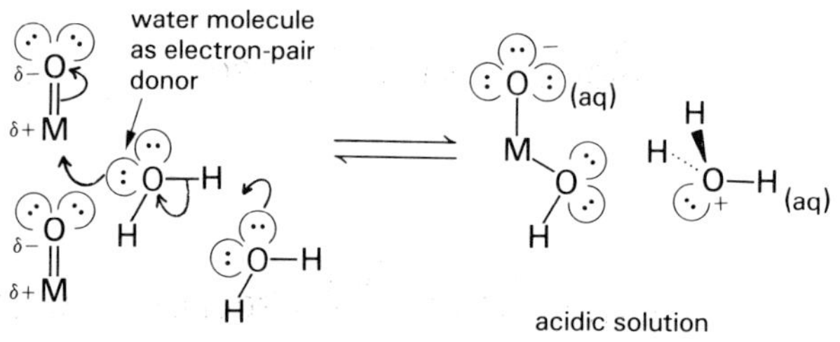

water molecule as electron-pair donor

acidic solution

In sodium oxide, the sodium ions hardly affect the lone pair donor properties of the oxide ions at all, and hence, sodium oxide is alkaline. In aluminium oxide, the degree of polarization is such that both effects are possible, and hence, the oxide is amphoteric and attacked by both alkali and acid. In sulphur trioxide, were it possible to have oxide ions, the lone pairs would be so extensively polarized that they would show no donor properties at all:

1 basic: $Na_2O(s) + H_2O(l) \longrightarrow 2Na^+(aq) + 2OH^-(aq)$

2 amphoteric (aluminium oxide is inert to water but the following reactions occur):

$Al_2O_3(s) + 6H_3O^+(aq) \longrightarrow 2Al^{3+}(aq) + 9H_2O(l)$

$Al_2O_3(s) + 2OH^-(aq) + 3H_2O(l) \longrightarrow 2Al(OH)_4^-(aq)$

3 acidic: $SO_3(s) + 2H_2O(l) \longrightarrow HSO_4^-(aq) + H_3O^+(aq)$

Non-metal sulphur compounds are mostly acidic. The hydride, oxides and halides are all hydrolysed to give acidic solutions, but carbon disulphide and the sulphur fluorides are inert.

$H_2S(g) + H_2O(l) \rightleftharpoons H_3O^+(aq) + HS^-(aq)$ $(pK_a = 7.05)$

$SO_2(g) + 2H_2O(l) \rightleftharpoons HSO_3^-(aq) + H_3O^+(aq)$ $(pK_a = 1.92)$

$SCl_4(l) + 8H_2O(l) \longrightarrow HSO_3^-(aq) + 4Cl^-(aq) + 5H_3O^+(aq)$

Redox properties

The elements themselves are oxidants, reflecting their tendency to combine in a negative oxidation state. Sulphur is a weaker oxidant, however, and can also act as a reductant because it combines in a positive oxidation state as well.

$O_2(g) + 4H_3O^+(aq) + 4e^- \rightleftharpoons 6H_2O(l)$ $E^\ominus = +1.23$ V   oxygen as oxidant

$S(s) + 2H_3O^+(aq) + 2e^- \rightleftharpoons H_2S(g) + 2H_2O(l)$ $E^\ominus = +0.14$ V   sulphur as oxidant

$H_2SO_3(aq) + 5H_3O^+(aq) + 4e^- \rightleftharpoons S(s) + 7H_2O(l)$ $E^\ominus = +0.45$ V   sulphur as reductant

Sulphur is neither a strong oxidant nor reductant as can be seen from the redox potentials. In fact, hydrogen sulphide reduces any sulphur oxyacid back to the elemental form. At the same time, hydrogen sulphide is oxidized to sulphur as well, for example:

$S(s) + 2H_3O^+_{(aq)} + 2e^- \rightleftharpoons H_2S(g) + 2H_2O(l) \;\|\; HSO_3^-(aq) + 5H_3O^+(aq) + 4e^- \rightleftharpoons S(s) + 8H_2O(l)$

$E^\ominus = +0.14$ V $E^\ominus = +0.45$ V

(−) ← electron flow → (+)

$2H_2S(g) + HSO_3^-(aq) + H_3O^+(aq) \longrightarrow 3S(s) + 4H_2O(l)$

Hydrogen peroxide and sulphur dioxide have an unusual redox property in common: they can both act as either an oxidant or a reductant, depending on the reaction conditions.

For example, either of these two compounds reduces manganate(VII), but oxidizes hydrogen sulphide.

$$5H_2O_2(aq) + 2MnO_4^-(aq) + 6H_3O^+(aq) \longrightarrow 2Mn^{2+}(aq) + 14H_2O(l) + 5O_2(g)$$

$$5SO_2(g) + 2MnO_4^-(aq) + 6H_2O(l) \longrightarrow 2Mn^{2+}(aq) + 4H_3O^+(aq) + 5SO_4^{2-}(aq)$$

Hydrogen sulphide is easily oxidized in solution, which further reflects the comparative lack of reactivity of elemental sulphur.

$$H_2S(g) + H_2O_2(aq) \longrightarrow S(s) + 2H_2O(l)$$

'Thio' compounds often have reducing properties as well. A 'thio' compound contains catenated sulphur oxyanions, often with a sulphur atom fulfilling a role usually taken by an oxygen atom. Some examples are shown on page 143.

thiosulphate	tetrathionate	dithionite
$S_2O_3^{2-}$	$S_4O_6^{2-}$	$S_2O_4^{2-}$

Sodium thiosulphate solution is used as a standard reducing reagent to analyse iodine solutions of unknown concentration:

$$I_2(aq) + 2S_2O_3^{2-}(aq) \longrightarrow 2I^-(aq) + S_4O_6^{2-}(aq)$$

Sodium dithionite is a very strong reductant and reacts quantitatively with oxygen in solution.

Sulphur(VI) has oxidizing properties. Concentrated sulphuric acid is not only a strong acid, it also acts as an oxidant and a dehydrating agent as well. The three different properties are illustrated by the reactions below.

1 As an oxidant (with copper or carbon):

$$\overset{0}{Cu}(s) + 2H_2\overset{VI}{S}O_4(l) \longrightarrow \overset{II}{Cu}SO_4(s) + \overset{IV}{S}O_2(g) + 2H_2O(l)$$

$$\overset{0}{C}(s) + 2H_2\overset{VI}{S}O_4(l) \longrightarrow \overset{IV}{C}O_2(g) + 2\overset{IV}{S}O_2(g) + 2H_2O(l)$$

2 As a dehydrant/oxidant (with glucose):

$$C_6H_{12}O_6(s) \xrightarrow[-6H_2O]{} C(s) \xrightarrow{as\ above} CO_2(g)$$

3 As an acid/dehydrant (with ethanedioates or methanoates):

[Structure of sodium ethanedioate] $\xrightarrow{2H^+}$ (COOH)$_2$ $\xrightarrow{-H_2O}$ CO + CO$_2$
ethanedioic acid

[Structure of sodium methanoate] $\xrightarrow{H^+}$ HCOOH $\xrightarrow{-H_2O}$ CO
methanoic acid

Solvent and complexing properties

Most solvents are either hydrocarbons like petrol and xylene, or compounds of oxygen like water H_2O, ethanol C_2H_5OH and propanone CH_3COCH_3. The oxygen solvents all tend to be polar because of the high electronegativity of oxygen. Sulphur, on the other hand, also forms solvents, but these resemble hydrocarbons more than they resemble the oxygen solvents. Liquids such as carbon disulphide CS_2 and dimethyl disulphide $CH_3-S-SCH_3$ are useful non-polar solvents because both carbon and sulphur have the same electronegativity. Many ligands (see page 120) co-ordinate through an oxygen or a sulphur atom. An example of each is shown below. The blood-red complex formed when thiocyanate ions are added to iron (III) is used as a test for iron (III).

ethanedioate ions as ligands (see page 243)	thiocyanate (SCN⁻) as ligands

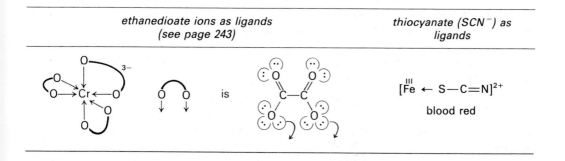

$[\overset{III}{Fe} \leftarrow S-C\equiv N]^{2+}$

blood red

9.3 GROUP V: NITROGEN AND PHOSPHORUS

Structure

Nitrogen and phosphorus atoms have five outer-shell electrons each. These are less tightly held to the nucleus than those of oxygen and sulphur respectively. Compare the electronegativities below.

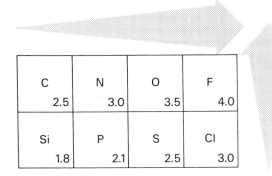

The same bonding features are found in the compounds of nitrogen and phosphorus as are found in those of oxygen and sulphur (see the discussion on page 136). The structure of the elements illustrates these features, nitrogen existing as $p\pi$-stabilized molecules while phosphorus adopts a σ-bonded arrangement.

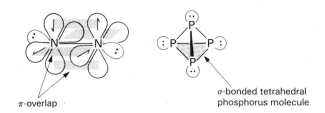

π-overlap

σ-bonded tetrahedral phosphorus molecule

Phosphorus exhibits monotropy (see page 52). The strained molecular geometry of a white phosphorus molecule accounts for the high reactivity of this form of the element. In other non-metallic groups, reactivity decreases going down the group, but white phosphorus is more reactive than nitrogen. The large bond energy of a nitrogen molecule contributes to the comparative inertness of nitrogen.

bond	bond energy/kJ mol^{-1}
N—N	163
N=N	409
N≡N	944

Compound formation

Like other non-metals, nitrogen and phosphorus combine with metals to produce compounds containing the non-metal in a negative oxidation state. These reactions take place only for the more reactive metals, and are less vigorous than those of the Group VI or VII elements with these metals.

$$6\text{Li}(s) + \overset{0}{\text{N}}_2(g) \xrightarrow{\text{heat}} 2\text{Li}_3\overset{-\text{III}}{\text{N}}(s)$$

$$6\text{Ca}(s) + \overset{0}{\text{P}}_4(s) \xrightarrow{\text{heat}} 2\text{Ca}_3\overset{-\text{III}}{\text{P}}_2(s)$$

The degree of covalency of nitrides and phosphides is high due to the large polarizability of the anions (see page 16). The chemistries of nitrogen and phosphorus are mostly concerned with positive oxidation states. Nitrogen forms a wide range of $p\pi$-stabilized oxycompounds as well as a huge variety of organo nitrogen compounds; phosphorus atoms make use of their vacant 3d-orbitals to form $d\pi$–$p\pi$ stabilized oxycompounds, as well as organophosphates and oxyhalides.

$p\pi$-stabilized nitrogen. A number of different oxidation states are available to nitrogen as a result of $p\pi$-bonding. The common oxides are:

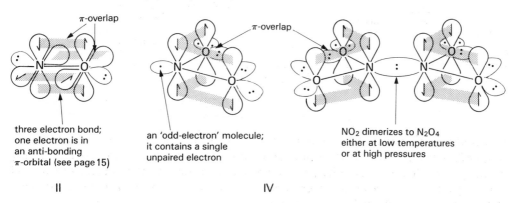

There are also several common oxyions e.g. NO_2^-, NO_2^+ and NO_3^-:

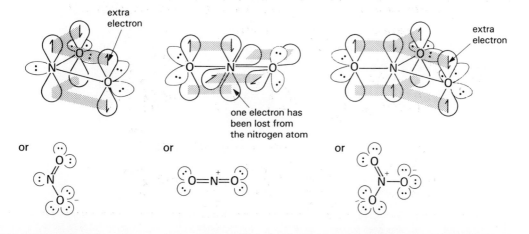

The II and IV state oxides are derived from concentrated nitric acid, which is itself the oxidation product of ammonia (see page 170).

$$\begin{cases} 4\overset{-III}{N}H_3(g) + 5O_2(g) \rightleftharpoons 4\overset{II}{N}O(g) + 6H_2O(g) \\ 4\overset{II}{N}O(g) + 2O_2(g) \longrightarrow 4\overset{IV}{N}O_2(g) \\ 4\overset{IV}{N}O_2(g) + O_2(g) + 2H_2O(l) \rightleftharpoons 4H\overset{V}{N}O_3(l) \end{cases}$$

$$3\overset{0}{Cu}(s) + 8H\overset{V}{N}O_3(l) \underset{50\% \text{ conc.}}{\longrightarrow} 3\overset{II}{Cu}(NO_3)_2(s) + 4H_2O(l) + 2\overset{II}{N}O(g)$$

$$\overset{0}{Cu}(s) + 4H\overset{V}{N}O_3(l) \underset{\text{conc.}}{\longrightarrow} \overset{II}{Cu}(NO_3)_2(s) + 2H_2O(l) + 2\overset{IV}{N}O_2(g)$$

Nitrogen(I) oxide can be prepared by decomposing ammonium nitrate(v).

$$\underset{\substack{\text{concentrated} \\ \text{aqueous solution}}}{\overset{-III}{N}H_4\overset{V}{N}O_3} \xrightarrow{\text{heat}} \overset{I}{N}_2O(g) + 2H_2O(g)$$

π-overlap

or

$$:N\equiv\overset{+}{N}-\overset{..}{\underset{..}{O}}:^-$$

dπ–pπ stabilized phosphorus. The π-overlap between a phosphorus 3p-orbital and an oxygen 2p-orbital is poor and so phosphorus forms no oxides analogous to the range of nitrogen oxides. However, by promoting one electron into the vacant, energetically low-lying 3d-orbitals, a phosphorus atom can combine to form a number of pπ–dπ stabilized oxyacids, as well as a V state oxide.

	phosphinic acid, $\overset{I}{H_3PO_2}$	*phosphonic acid,* $\overset{III}{H_3PO_3}$	*orthophosphoric acid,* $\overset{V}{H_3PO_4}$
	3d$_{xy}$ orbital, π-overlap, 2p$_y$ orbital		
	or	or	or
	HO\P=O / H H	HO\P=O / H OH	HO\P=O / HO OH

9 Chemistry of the Non-metals

The I state acid (phosphinic acid) is produced with phosphine during the disproportionation of white phosphorus in boiling alkali.

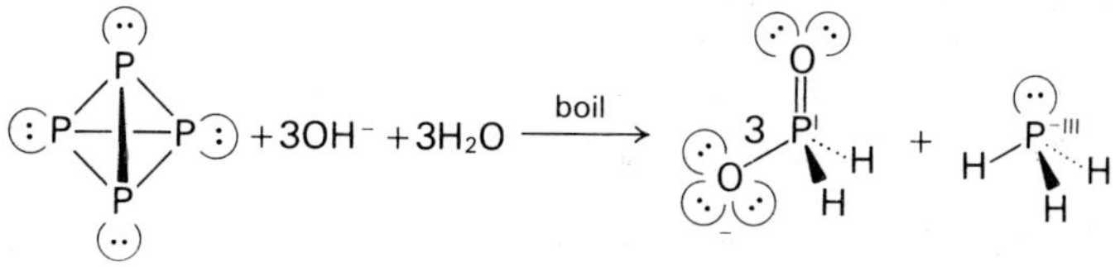

$$P_4 + 3OH^- + 3H_2O \xrightarrow{boil} 3 H_2PO_2^- + PH_3$$

(Notice that other non-metals also disproportionate in alkali: (see pages 132 and 138.) The III state and V state acids (phosphonic acid and phosphoric acids) are produced by hydrolysis of the corresponding oxides or halides. For example, when phosphorus burns in excess oxygen, the following reaction occurs.

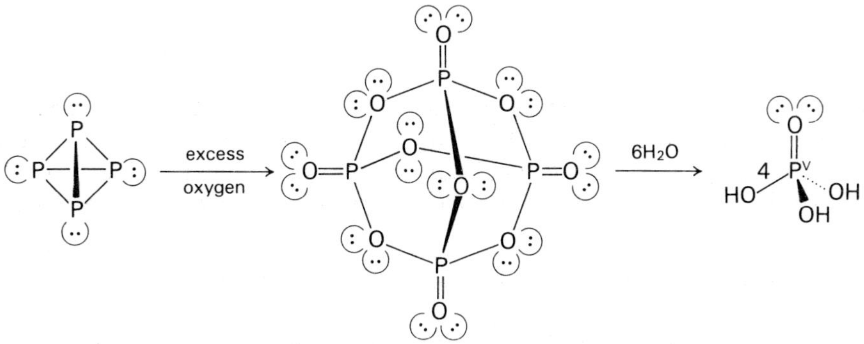

$$P_4 \xrightarrow{excess~oxygen} P_4O_{10} \xrightarrow{6H_2O} 4\,H_3PO_4$$

σ-bonded nitrogen and phosphorus. Both nitrogen and phosphorus form hydrides and halides that are trivalent and contain only σ-bonds. Phosphorus(III) oxide and phosphorus(V) chloride also contain σ-bonds throughout. All four (shown below) can be prepared by direct combination.

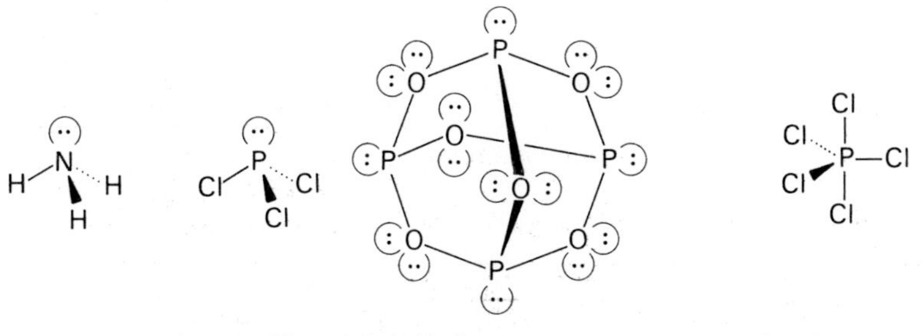

NH_3 PCl_3 P_4O_6 PCl_5

prepared by reacting chlorine or oxygen with excess phosphorus   phosphorus and excess chlorine

A very low yield of ammonia results from the reaction of nitrogen and hydrogen (see page 169), but the above two phosphorus(III) compounds are produced in good yield from their elements providing excess phosphorus is used. With excess chlorine or oxygen, the V state compounds are synthesized.

Acid-base properties

Metal nitrides and phosphides are strongly basic and dissolve to give alkaline solutions. Ammonia (or phosphine) is produced.

$$Li_3N(s) + 3H_2O(l) \longrightarrow 3Li^+(aq) + 3OH^-(aq) + NH_3(g)$$
$$Ca_3P_2(s) + 6H_2O(l) \longrightarrow 3Ca(OH)_2(s) + 2PH_3(g)$$

Ammonia is also weakly alkaline because of the lone pair donor properties of an ammonia molecule.

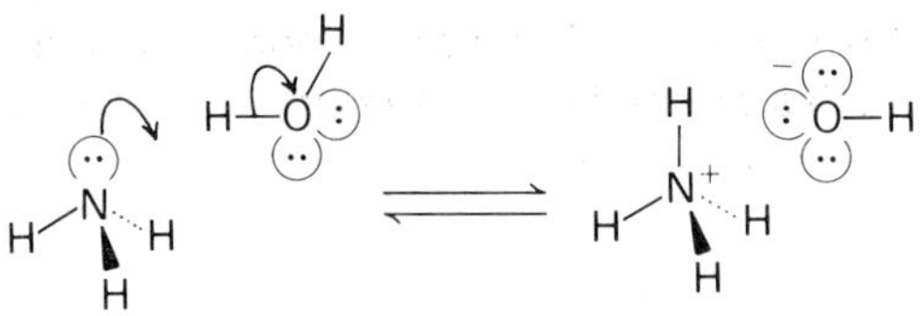

$$K_b = 10^{-4.8} \text{ mol dm}^{-3}$$

Phosphine, however, shows almost no basic properties at all: it is neutral and insoluble in water and the phosphonium ion (PH_4^+) has an extremely limited chemistry compared with that of the ammonium ion. A huge number of ammonium salts exist, both simple ones such as chloride and sulphate as well as double salts such as those of iron:

$$\overset{II}{Fe}(NH_4)_2(SO_4)_2 \cdot 6H_2O \quad \text{and} \quad \overset{III}{Fe}(NH_4)(SO_4)_2 \cdot 12H_2O$$

The lack of basic character associated with phosphine is due to the weak phosphorus–hydrogen bond and the small lattice or hydration energies resulting from the formation of phosphonium ions. Although the lone pair of electrons on a phosphine molecule is prominent, it is rather diffuse because of the comparatively low electronic control of a phosphorus atom (electronegativity = 2.1). Compare the molecular geometry and energy factors involved in the interaction of the two hydrides with water (as shown above for ammonia).

	energy factors	
bond forming	bond energy/kJ mol^{-1}	hydration energy/kJ mol^{-1} of the cation formed
N—H	−388	−301
P—H	−322	−225

	molecular geometry	
bonds		bond angle/degrees
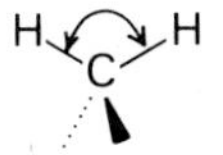 methane (H–C–H)		109.5
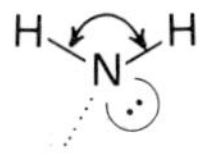 ammonia (H–N–H)		107.0
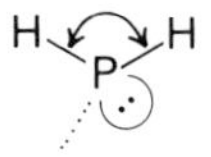 phosphine (H–P–H)		93.0

In positive oxidation states, nitrogen and phosphorus form compounds that are mostly acidic in solution. Dinitrogen oxide and nitrogen monoxide are exceptions and are neutral. Nitrogen dioxide disproportionates in water to give nitrate(III) (nitrite) and nitrate(V).

$$2\overset{IV}{N}O_2(g) + 2H_2O(l) \rightleftharpoons H\overset{III}{N}O_2(aq) + \overset{V}{N}O_3^-(aq) + H_3O^+(aq)$$

The phosphorus oxides and halides give phosphonic or phosphoric acid depending on the original phosphorus oxidation state. It is interesting to note that the three phosphorus acids shown below all have the same strength because the molecules of each have only one double-bonded oxygen atom (see the discussion of acid strength on page 161): $pK_a \approx 2$ for each.

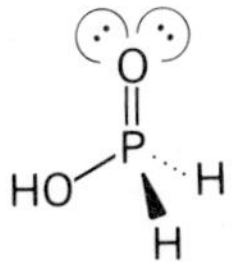

monobasic 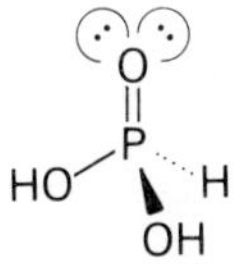   dibasic 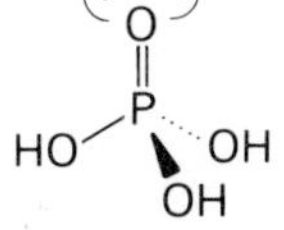   tribasic

Nitrogen trichloride is not hydrolysed in the same way as phosphorus trichloride; nitrogen and chlorine are equally electronegative, and so the terminal chlorine atoms are attacked instead of the central atom. A chlorine atom can accommodate another electron pair in its outer shell, whereas the outer shell of nitrogen is limited to eight.

When the above step is repeated for the other two chlorine atoms bonded to the nitrogen atom, the following equation summarizes the change.

$$NCl_3(l) + 3H_2O(l) \longrightarrow NH_4^+(aq) + ClO^-(aq) + 2HOCl(aq)$$
weak acid

Redox chemistry

Reductants: nitrogen(−III) and phosphorus(−III) both having reducing properties. For example, ammonia reduces copper(II) oxide in the solid state while phosphine, a stronger reductant, reduces silver(I) in aqueous solution.

$$2\overset{II}{Cu}O(s) + 2\overset{-III}{N}H_3(g) \longrightarrow 2\overset{0}{Cu}(s) + \overset{0}{N}_2(g) + 3H_2O(g)$$

$$8\overset{I}{Ag}^+(aq) + \overset{-III}{P}H_3(g) + 12H_2O(l) \longrightarrow 8\overset{0}{Ag}(s) + H_3\overset{V}{P}O_4(aq) + 12H_3O^+(aq)$$

The catalytic oxidation of ammonia to nitric acid is discussed on page 170. Nitrogen(II) and nitrogen(IV) also act as reductants, although there are few uses of these redox couples. Nitrogen monoxide is colourless, but is very rapidly air-oxidized to brown nitrogen. dioxide:

$$2\overset{II}{N}O(g) + \overset{0}{O}_2(g) \longrightarrow 2\overset{IV}{N}O_2(g)$$

Phosphorus(I) and (III) are powerfully reducing. A solution of sodium phosphinate is a useful reductant to carry out the reduction of diazonium ions in alkali (page 209).

$$2\,C_6H_5N_2^+(aq) + 2OH^-(aq) + 2H_2PO_2^-(aq) \longrightarrow 2\,C_6H_6(l) + H_2PO_4^-(aq) + N_2(g)$$

Oxidants: nitrogen(v) is considerably less stable than phosphorus(v) and this is reflected by two properties in particular. Firstly, nitrate(v) salts tend to decompose to give compounds containing nitrogen in a lower oxidation state.

$$2Na\overset{V}{N}O_3(s) \xrightarrow{heat} 2Na\overset{III}{N}O_2(s) + O_2(g)$$

$$2Pb(\overset{V}{N}O_3)_2(s) \xrightarrow{heat} 2PbO(s) + 4\overset{IV}{N}O_2(g) + O_2(g)$$

Phosphate(v) salts do not tend to decompose in the same way; instead, there are many condensed forms of phosphorus(v), for example, sodium trimetaphosphate $Na_3P_3O_9$.

$$3Na_2H\overset{V}{P}O_4(s) \xrightarrow{heat} Na_3\overset{V}{P}_3O_9(s) + 3H_2O(g)$$

Secondly, concentrated nitric acid is a strong oxidant, whereas phosphoric acid has little oxidizing character. Nitric acid oxidizes copper (page 146), carbon and many organic compounds, amongst others. Compare the redox potentials given below.

$$\overset{V}{N}O_3^-(aq) + 3H_3O^+(aq) + 2e^- \rightleftharpoons H\overset{III}{N}O_2(aq) + 4H_2O(l) \qquad E^\ominus = +0.94 \text{ V}$$

$$H_3\overset{V}{P}O_4(aq) + 2H_3O^+(aq) + 2e^- \rightleftharpoons H_3\overset{III}{P}O_3(aq) + 3H_2O(l) \qquad E^\ominus = -0.28 \text{ V}$$

Hence, nitrate(v) and phosphate(III) react to produce nitrate(III) (also some nitrogen oxides) and phosphate(v).

$$H_3\overset{III}{P}O_3(aq) + \overset{V}{N}O_3^-(aq) + H_3O^+(aq) \longrightarrow H_3\overset{V}{P}O_4(aq) + H\overset{III}{N}O_2(aq)$$

Complexing properties

Ammonia and its organic derivatives (amines) act as complexing agents. The lone pair donor properties of the molecules lead them to behave as ligands. Many examples of this are described in the chemistry of the metals (see, for example, page 120). The phenyl derivatives of phosphine stabilize transition metals in low oxidation states but, otherwise, phosphine has a less extensive complex chemistry than ammonia.

Nitrogen monoxide molecules act as ligands to form a brown coloured complex of iron(II) that is produced in the 'brown ring' test for nitrates. When iron(II) sulphate is added to a nitrate solution, and concentrated sulphuric acid is poured carefully into the system to form a separate lower layer, a brown ring develops at the boundary. Under the non-standard conditions, nitrogen(v) is reduced by iron(II) to nitrogen monoxide which then complexes with unreacted iron(II). The whole reaction takes place at the acid surface. The complex has the formula $[\overset{II}{Fe}(H_2O)_5NO]^{2+}$.

9.4 GROUP IV

Structure

The Group IV atoms have full inner s, p and d-subshells, and an outer-shell configuration ns^2np^2. When all the outer-shell electrons are used for bonding, the compounds formed have high degrees of covalency. There is, however, an increased tendency for the ns^2 electrons to be withdrawn into the atomic core. This is known as the inert pair effect and is described on page 126. The values of the electron affinities, ionization energies and electronegativities are all in the middle of the range: high for metals but low for non-metals.

The physical structure of the Group IV elements neatly illustrates these generalizations. Carbon exhibits monotropy in forming two macromolecular structures in which each carbon atom is tetra-covalent. Silicon and germanium form distorted diamond-type lattices with semi-metal characteristics (both are semi-conductors). Tin exhibits enantiotropy: grey tin has a similar structure to that of germanium but, above 14°C this collapses into a close-packed metal structure of divalent cations and delocalized electrons

diamond structure	*graphite structure*	
$d_1 = 0.154$ nm	$d_2 = 0.142$ nm	$d_3 = 0.335$ nm

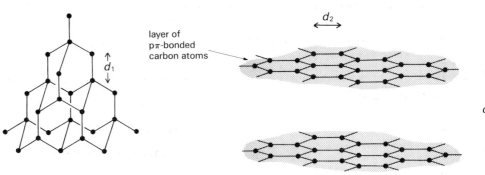

diamond is less stable than graphite by about 5 kJ mol^{-1}. The C—C bond angle is smaller in the diamond lattice and therefore there is a greater degree of repulsion between the bonding pairs.

each pπ-bonded layer contains the interlinking hexagons shown below:

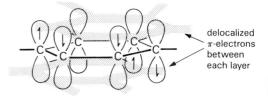

delocalized π-electrons between each layer

(white tin). Lead exists only as a metallic lattice. The changes reflect the fall-off in electronic control exerted by a Group IV atom as the atomic number increases. The monotropes of carbon are graphite and diamond. The bonding in graphite illustrates further the tendency of the Period two non-metals to form pπ-stabilized structure:

Compound formation

The steady emergence of metallic character going down the group is evident in the range of compounds formed. In keeping with its position as a second Period non-metal, carbon is found in a huge number of pπ-stabilized compounds. The other Group IV elements form no compounds analogous to these carbon compounds. For the remainder, the inert pair effect becomes increasingly dominant and, therefore, the stability of the II state increases.

pπ-stabilized carbon is present in simple oxides such as CO and CO_2, but is also responsible for a vast range of organic molecules: alkenes, ketones, acids, amides, aromatic systems and nitriles. The examples are too numerous to list but a few are drawn below.

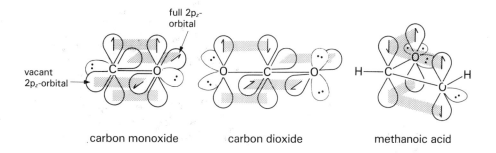

carbon monoxide   carbon dioxide   methanoic acid

The dicarbide ion is a molecular anion with bonding features like those shown above. A dicarbide is produced when carbon is heated to about 2000°C with a reactive metal oxide. For example,

$$CaO(s) + 3C(s) \xrightarrow{heat} CaC_2(s) + CO(g)$$

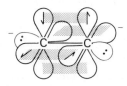

Dicarbide ions are isoelectronic with nitrogen molecules and are readily protonated to form ethyne.

$$CaC_2(s) + 2H_2O(l) \longrightarrow Ca(OH)_2(s) + C_2H_2(g)$$

C_2^{2-} ions have a high polarizability (like carbonate ions, page 114), and are not formed in the presence of small highly-charged cations. Instead, carbide ions (C^{4-}) are generated and the resulting lattice then has an even higher degree of covalency than a dicarbide lattice. Carbides of this sort are also hydrolysed differently; for example, methane is produced from aluminium carbide.

$$Al_4C_3(s) + 12H_3O^+(aq) \xrightarrow{boil} 4Al^{3+}(aq) + 12H_2O(l) + 3CH_4(g)$$

The IV state. Tetravalent hydrides, halides and oxides of all the elements are well known. These contain the elements in oxidation state IV (except CH_4), and are the result of the atoms using all their valence electrons for bonding. The halides and oxides can be prepared by direct combination and, with the exception of $p\pi$-stabilized CO_2, have similar structures:

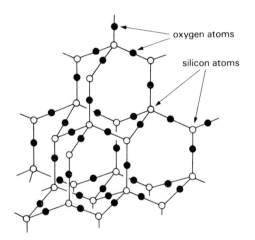

although there is charge separation along each bond, there is no dipole moment

'diamond' structure of SiO_2 (quartz) lattice; the other MO_2 lattices are distorted versions of this

The hydrides (except methane) are generated from the chlorides using lithium tetrahydridoaluminate. They are rather explosive compounds that are inflammable and are also easily hydrolysed.

$$MCl_4(l) + LiAlH_4(s) \longrightarrow LiCl(s) + AlCl_3(s) + MH_4(g)$$

The II state. Compounds of the elements in the II state include oxides, halides and salts derived from various acids. Lead has the most extensive II state chemistry, because the inert pair effect is most marked for lead atoms. However, germanium(II) oxide can be produced by heating the dioxide and the element strongly together.

$$\overset{0}{Ge}(s) + \overset{IV}{GeO_2}(s) \rightleftharpoons 2\overset{II}{Ge}O(s)$$

Tin(II) is also well characterized as an oxide, a chloride (which appears to be molecular and remains appreciably associated even under aqueous conditions) and in a number of complexes described later. Tin and lead show typical metallic properties in their (weak) ability to displace hydrogen from an aqueous acid.

$$Pb(s) + 2H_3O^+(aq) \xrightarrow{heat} Pb^{2+}(aq) + 2H_2O(l) + H_2(g)$$

Compare their electrode potentials with that of a more reactive B-metal, zinc:

$$Zn^{2+}(aq) + 2e^- \rightleftharpoons Zn(s) \quad E^\ominus = -0.76 \text{ V}$$
$$Sn^{2+}(aq) + 2e^- \rightleftharpoons Sn(s) \quad E^\ominus = -0.14 \text{ V}$$
$$Pb^{2+}(aq) + 2e^- \rightleftharpoons Pb(s) \quad E^\ominus = -0.13 \text{ V}$$

Catenated compounds. The tetravalency of Group IV elements and the comparatively high energy of the bonds produced explain their tendency to form catenated compounds (compounds whose molecules contain atoms of the same element bonded into rings or chains onto which other atoms can be attached; Latin *catena* = chain). Carbon has an immense chemistry illustrating this property (organic chemistry), but silanes and even germanes can also be synthesized. These are less stable because of the lower bond energies, but even so, a compound of formula Si_6H_{14} has been isolated from the hydrolysis products of magnesium silicide.

$$2Mg(s) + Si(s) \xrightarrow[heat]{strong} Mg_2Si(s) \xrightarrow[boil]{H_3O^+} SiH_4, Si_2H_6 \ldots Si_6H_{14}(g)$$

Acid-base properties

The oxides and hydroxyl compounds. In keeping with the trend of increasing metallic character, the oxides and hydroxyl compounds of the Group IV elements change from acids through to amphoteric bases down the group.

1 Acidic:

$$CO_2(g) + 2H_2O(l) \rightleftharpoons HCO_3^-(aq) + H_3O^+(aq)$$

2 Amphoteric:

$$Sn(OH)_2(s) + 2H_3O^+(aq) \longrightarrow Sn^{2+}(aq) + 4H_2O(l)$$
$$Sn(OH)_2(s) + 2OH^-(aq) \longrightarrow Sn(OH)_4^{2-}(aq)$$

3 Basic:

$$PbO(s) + 2H_3O^+(aq) \longrightarrow Pb^{2+}(aq) + 3H_2O(l)$$

Silicon dioxide (silica) is an acidic oxide that occurs as an impurity in the iron ore added to the blast furnace. Basic calcium oxide is added to convert the silica to a silicate 'slag'.

$$\underset{acid}{SiO_2(s)} + \underset{base}{CaO(s)} \xrightarrow{heat} \underset{salt\ (slag)}{CaSiO_3(s)}$$

An enormous number of different silicate structures exist. These are all derived from the acidic oxide silica, which can be thought of as a lone pair acceptor (Lewis acid):

$$SiO_2 + 2\,\ddot{O}^{2-} \longrightarrow [SiO_4]^{4-}$$

The linking between the tetrahedral [SiO$_4$] units is varied and complex. If each is linked to four other units, the lattice becomes that of pure silica; if there are 'unlinked' oxygen atoms, metal cations then complete the lattice. Many rocks, clays and micas are silicate structures of this sort.

Halides and hydrides. With the exception of those of carbon, the tetrahalides and hydrides are hydrolysed in water to give acidic solutions. The tetrafluorides, however, are more resistant to hydrolysis.

$$SiH_4(g) + 4H_2O(l) \longrightarrow SiO_2 \cdot 2H_2O(s) + 4H_2(g)$$

$$SnCl_4(l) + 8H_2O(l) \longrightarrow Sn(OH)_4(s) + 4H_3O^+(aq) + 4Cl^-(aq)$$

The inertness of carbon tetrachloride (tetrachloromethane) to hydrolysis, even under alkaline conditions, can be attributed to two factors.

1. The attacking nucleophile is sterically hindered (see page 245) because there are four bulky chlorine atoms on one small carbon atom.
2. The carbon atom cannot accept another pair of electrons into its outer shell (unlike any other Group IV atom which has a vacant outer d-subshell). This makes electron-pair donation even less likely.

Redox properties

The 0 state. Carbon and silicon are used to reduce metal ores. Carbon is the reductant for haematite in the blast furnace.

$$3\overset{0}{C}(s) + \overset{III}{Fe_2}O_3(s) \xrightarrow{heat} 2\overset{0}{Fe}(l) + 3\overset{II}{C}O(g)$$

Silicon is used to reduce magnesium oxide to magnesium. This process is carried out at very low pressure so that the magnesium is extracted as a vapour from above the molten metal produced.

$$2\overset{II}{Mg}O(s) + \overset{0}{Si}(s) \xrightarrow{heat} 2\overset{0}{Mg}(g) + \overset{IV}{Si}O_2(s)$$

Tin and lead are weak metallic reductants reflecting their low reactivity as metals. For example, lead displaces copper slowly from a solution of copper sulphate, but is displaced by zinc from a solution of lead(II) nitrate.

$$Pb(s) + Cu^{2+}(aq) \longrightarrow Pb^{2+}(aq) + Cu(s)$$

$$Zn(s) + Pb^{2+}(aq) \longrightarrow Zn^{2+}(aq) + Pb(s)$$

The II state. As described earlier, the inert pair effect becomes more dominant at the foot of the group and, therefore, the II state becomes increasingly preferred. Hence, Ge(II) and Sn(II) are strong reductants but Pb(II) is not. Compare the redox potentials shown below.

$$Sn(IV) + 2e^- \rightleftharpoons Sn(II) \quad E^\ominus = +0.15 \text{ V}$$
$$Pb(IV) + 2e^- \rightleftharpoons Pb(II) \quad E^\ominus = +1.69 \text{ V}$$

Aqueous tin(II) is a useful laboratory reductant; for example, iron(III) is reduced to iron(II) and iodine(0) to iodide(−I). Tin(II) chloride and concentrated hydrochloric acid are the reagents used to reduce nitrobenzene to phenylamine (see page 207).

The chemistry of carbon(II) does not follow the outlines described above. Carbon(II) is almost invariably $p\pi$-stabilized as in carbon monoxide. Nonetheless, there is a tendency for carbon(IV) to be thermodynamically more stable so that carbon monoxide acts as a reductant. For example, the carbon monoxide/carbon dioxide couple is vital in metal extraction (see page 166). Also carbon monoxide undergoes gaseous redox reactions such as that with chlorine.

$$\overset{II}{C}O(g) + \overset{0}{Cl_2}(g) \xrightarrow[\text{charcoal}]{\text{activated}} \overset{IV}{C}O\overset{-I}{Cl_2}(g)$$

It disproportionates when heated strongly.

$$2\overset{II}{C}O(g) \rightleftharpoons \overset{0}{C}(s) + \overset{IV}{C}O_2(g)$$

The IV state. The trend in stability of the II state is reversed when the IV state is considered. Ge(IV) has few redox properties, but Pb(IV) is quite a strong oxidant. For example, lead(IV) oxide oxidizes concentrated hydrochloric acid to chlorine.

$$\overset{IV}{P}bO_2(s) + 4\overset{-I}{H}Cl(\text{conc}) \longrightarrow \overset{II}{P}bCl_2(s) + 2H_2O(l) + \overset{0}{Cl_2}(g)$$

It decomposes on heating to give lead(II) oxide.

$$2\overset{IV}{P}bO_2(s) \longrightarrow 2\overset{II}{P}bO(s) + O_2(g)$$

As before, the chemistry of carbon(IV) is anomalous. Carbon dioxide acts as an oxidant in its reaction with red-hot carbon or with molten magnesium.

$$\overset{IV}{C}O_2(g) + \overset{0}{C}(s) \xrightarrow[\text{heat}]{\text{strong}} 2\overset{II}{C}O(g)$$

$$\overset{IV}{C}O_2(g) + \overset{0}{M}g(s) \xrightarrow[\text{heat}]{\text{strong}} \overset{II}{M}gO(s) + \overset{II}{C}O(g)$$

Complexing properties

Electron-pair acceptors. With the exception of carbon, compounds of the elements in both II and IV state form complexes. The vacant outer d-subshell is used by the bonded atoms to accept the electron-pairs of the ligands. Carbon compounds cannot form analogous complexes because there are no d-orbitals in the outer shell of a carbon atom.

Some typical examples of complex formation are shown below. Note that both tin(II) and lead(II) are amphoteric because of the formation of hydroxyl complexes.

$$SiO_2(s) + 6HF(conc) \longrightarrow SiF_6^{2-}(aq) + 2H_3O^+(aq)$$
$$Pb(OH)_2(s) + 2OH^-(aq) \longrightarrow Pb(OH)_4^{2-}(aq)$$

Lead(II) chloride is insoluble in cold water, but dissolves in concentrated hydrochloric acid because of complex formation. This reaction is used in analysis to detect lead(II) ions.

$$PbCl_2(s) + 2Cl^-(aq) \longrightarrow PbCl_4^{2-}(aq)$$

Electron-pair donors. Carbon monoxide molecules and cyanide ions are both ligands that contain $p\pi$-bonded carbon atoms. As ligands they tend to stabilize transition metals in low oxidation states because they are σ-donors but π-acceptors (see page 120). For example, carbon monoxide is used in the purification of nickel: it combines with the metal to form a carbonyl from which nickel can readily be extracted.

$$Ni(s) + 4CO(g) \underset{high\ T}{\overset{low\ T}{\rightleftharpoons}} Ni(CO)_4(g)$$

Cyanide ions complex an enormous range of metal cations. For example, potassium hexacyanoferrate(II) is a laboratory reagent which gives a Prussian blue precipitate when added to iron(III) ions.

$$Fe^{3+}(aq) + K^+(aq) + Fe(CN)_6^{4-}(aq) \longrightarrow \underset{\text{Prussian blue}}{KFe[Fe(CN)_6](s)}$$

Thiocyanate ions (CNS$^-$) are also used in the analysis of iron(III) because they produce a blood-red complex of formula $Fe(CNS)^{2+}_{(aq)}$.

9.5 HYDROGEN

Structure

A hydrogen atom has the simplest structure of all atoms. It has a nucleus containing one proton and a single electron in a 1s-orbital. There are three isotopic forms of hydrogen containing respectively zero, one and two neutrons per atom: protium ^1_1H, deuterium ^2_1H and tritium ^3_1H. The relative atomic mass of hydrogen is 1.008 and reflects the 0.015% presence of deuterium in a sample of hydrogen. There is only a tiny trace of tritium: it is radioactive and undergoes β-decay with a half-life of 12.3 years. Hydrogen gas contains diatomic H_2 molecules. The valence shell is full when there are two electrons in it.

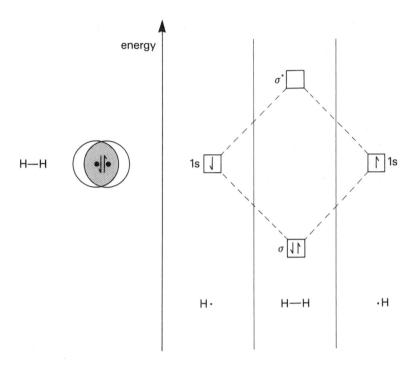

Compound formation

The electronegativity of hydrogen is 2.1 and it is, therefore, among the least electronegative of non-metals. A hydrogen atom has only one valence electron, so the chemistry of the element is similar in some ways to that of a Group I element; however, because only one more electron is required to complete the valence shell, its chemistry also resembles that of a halogen an occasion. Hydrogen forms three different classes of compound illustrating this versatility.

Compounds with non-metals. The high bond dissociation energy of a hydrogen molecule (436 kJ mol^{-1}) makes the element appear less reactive than it otherwise might be. It combines directly with many non-metals to form molecular hydrides; the violence of the reactions varies with the reactivity of the non-metal. For example,

$H_2(g) + F_2(g) \longrightarrow 2HF(g)$ explosive (even in the dark)
$H_2(g) + I_2(g) \rightleftharpoons 2HI(g)$ $K_p = 4.6$ at 500°C and 0.1 MPa
$3H_2(g) + N_2(g) \rightleftharpoons 2NH_3(g)$ $K_p = 1.6 \times 10^{-7}$ MPa^{-2} at 500°C and 0.1 MPa

There is charge separation along the bonds of these hydride molecules, and for the hydrides of nitrogen, oxygen and fluorine this leads to hydrogen bonding (see page 20) between the molecules. For example, water and hydrogen fluoride have anomalously high melting and boiling points for a substance whose molecules have only ten electrons each. The solid lattice of hydrogen fluoride contains zig-zag chains of hydrogen-bonded molecules as shown overleaf.

(i)

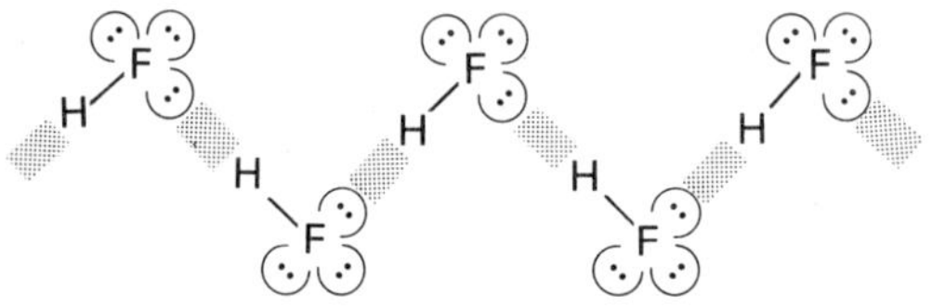

T_m −83 °C
T_b 20 °C

Compounds with reactive metals. Hydrogen combines with certain s-block metals to form salt-like hydrides. These are ionic in character but have appreciable degrees of covalency because a hydride ion has a fairly high polarizability (its size is between that of a bromide and an iodide ion).

$$2Na(s) + H_2(g) \longrightarrow 2NaH(s)$$

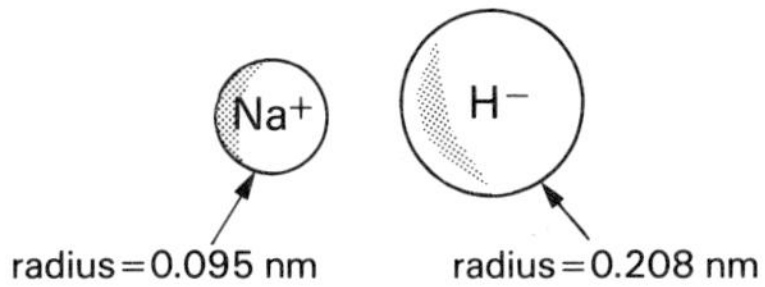

Na+ radius = 0.095 nm
H− radius = 0.208 nm

Interstitial compounds with transition metals. Hydrogen forms a range of non-stoichiometric compounds with the more electron-rich transition metals. In particular, the nickel, palladium and platinum group show a great affinity for hydrogen. The non-metal is absorbed into the metal lattice in an atomic state. The half-full 1s-orbitals of the hydrogen atoms overlap with the d-orbitals of the metal atoms.

Hydrogen also combines with the element boron to produce a number of unusual electron-deficient molecules. A typical example is diborane (B_2H_6) which was first isolated when magnesium boride was hydrolysed by phosphoric acid. A number of other boranes (e.g. B_4H_{10}) are also formed during the reaction whose equation is given below.

$$Mg_3B_2(s) + 3H_3PO_4(l) \xrightarrow{warm} 3MgHPO_4(s) + B_2H_6(g)$$

The electronic structure of a borane contains three-centred bonds that have the appearance of an ordinary σ-bond with a proton trapped in the centre.

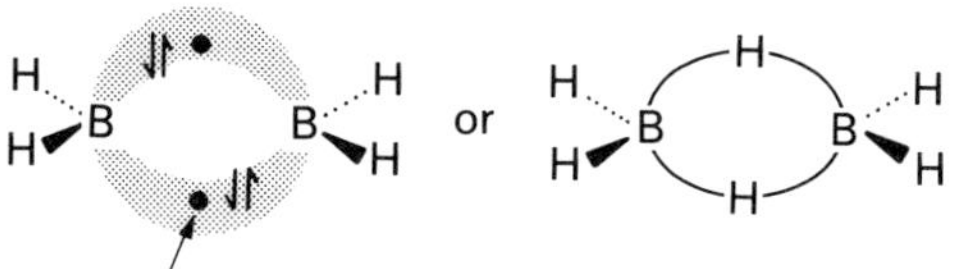

trapped proton

Acid-base properties

The acid-base theory of Brønsted and Lowry is concerned with the transfer of protons between particles (see page 84). The proton of a hydrogen nucleus is the only proton not buried deep inside an atomic core. When a hydrogen atom bonds to a more electronegative atom, the proton can be attracted to, and accepted by, a second particle with a lone pair of electrons. A water molecule, for example, has two electron-deficient protons and also two lone pairs of electrons as well. Water is amphoteric:

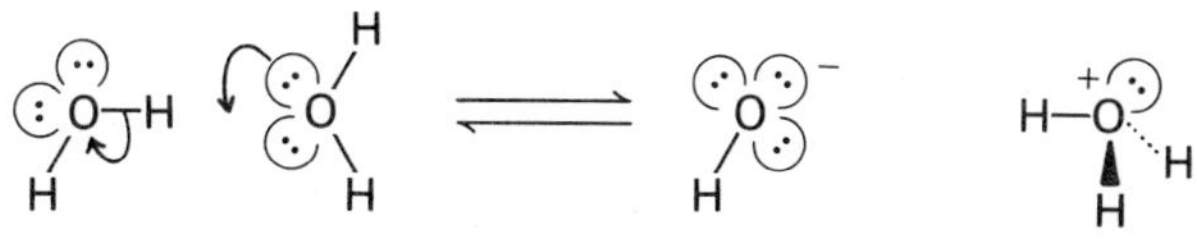

base   acid

$$^{298}K_w = [H_3O^+][OH^-] = 10^{-14} \text{ mol}^2 \text{ dm}^{-6}$$

In pure water at 298 K, $[H_3O^+] = [HO^-] = 10^{-7}$ mol dm^{-3}. Any substance that increases the concentration of hydroxonium ions is acidic; any that increases the concentration of hydroxide ions is alkaline.

Acids. Acidic substances fall into three main categories:

1 non-metal hydrides, for example the hydrides of the halogens;

$$X-H(g) + H_2O(l) \longrightarrow X^-(aq) + H_3O^+(aq)$$

2 the salts of metals whose cations have a high polarizing power;

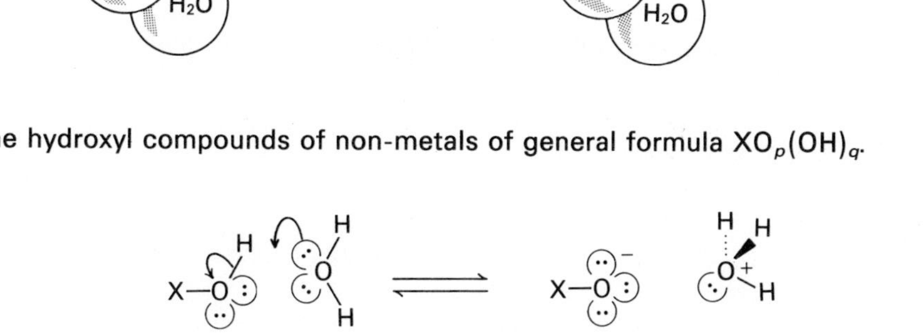

3 the hydroxyl compounds of non-metals of general formula $XO_p(OH)_q$.

Most of the common acids belong to the third category (oxyacids). For example, consider the following acids and their respective strengths shown on page 162.

acid	formula	XO$_p$(OH)$_q$	p	q	K_a/mol dm^{-3}
sulphuric(VI)	H$_2$SO$_4$	SO$_2$(OH)$_2$	2	2	~10^3
sulphurous(IV)	H$_2$SO$_3$	SO(OH)$_2$	1	2	10$^{-1.9}$
chloric(VII)	HClO$_4$	ClO$_3$(OH)	3	1	~10^{10}
chloric(V)	HClO$_3$	ClO$_2$(OH)	2	1	~10^2
chloric(III)	HClO$_2$	ClO(OH)	1	1	10$^{-2.0}$
chloric(I)	HClO	Cl(OH)	0	1	10$^{-7.4}$
bromic(I) acid	HBrO	Br(OH)	0	1	10$^{-8.7}$
phosphoric(V)	H$_3$PO$_4$	PO(OH)$_3$	1	3	10$^{-2.1}$
phosphonic(III)	H$_3$PO$_3$	HPO(OH)$_2$	1	2	10$^{-2.0}$
phosphinic(I)	H$_3$PO$_2$	H$_2$PO(OH)	1	1	10$^{-2.1}$

From the data, it can be seen that p, the number of double-bonded oxygen atoms, appears to be the dominant factor controlling acid strength. The number of hydroxyl groups does not seem to affect the acid strength. As p increases, the stability of the acid anion increases because there are more ways by which the charge can be delocalized. For example, in a chlorate(VII) ion:

The more stable the acid anion, the further to the right the position of the acid equilibrium shifts and, therefore, the stronger is the acid.

$$HA(aq) + H_2O(l) \rightleftharpoons H_3O^+(aq) + A^-(aq) \qquad K_a \text{ is large}$$

The electronegativity of the central atom also has some effect. For example, the acid strength of the simple hydroxyhalides (halogen(I) acids) decreases as the electronegativity of the halogen decreases. This is because a more electronegative atom stabilizes the negative charge of the anion better than a less electronegative atom.

Alkalis. There are three types of hydrogen compounds that are alkaline;

1 ammonia:

$$NH_3(g) + H_2O(l) \rightleftharpoons NH_4^+(aq) + OH^-(aq)$$

2 metal hydrides:

for example,

$$NaH(s) + H_2O(l) \longrightarrow Na^+(aq) + OH^-(aq) + H_2(g)$$

3 metal hydroxides:

$$KOH(s) + (aq) \longrightarrow K^+(aq) + OH^-(aq)$$

All metal hydroxides are basic, but only a few are alkaline (because most are insoluble). However, the Group I hydroxides *are* soluble, and sodium hydroxide is often known simply as 'alkali' for this reason. It is instructive to compare and contrast the properties of the hydroxyl compounds of the elements going across the third period. Two clear trends are evident:

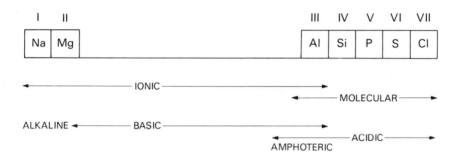

See the discussion of oxides on page 140. An oxide can be viewed as the anhydride of its corresponding hydroxyl compound.

Redox properties

Hydrogen(0). Elemental hydrogen acts as a reductant in its reaction with non-metals or with the oxides of unreactive metals.

$$\overset{0}{N_2}(g) + 3\overset{0}{H_2}(g) \rightleftharpoons 2\overset{-III\ I}{NH_3}(g)$$

$$\overset{II}{Cu}O(s) + \overset{0}{H_2}(g) \xrightarrow{heat} \overset{0}{Cu}(s) + \overset{I}{H_2}O(g)$$

Molecular hydrogen is not as strong a reductant as the atomic form found in the lattice of an interstitial transition metal hydride (see page 192). In this form, hydrogen is still in the zero state because there is almost no charge transferred between the hydrogen and transition metal atoms. For example, hydrogen in the presence of nickel or palladium is used to reduce unsaturated organic compounds such as alkenes and nitriles.

$$\underset{H}{\overset{R}{>}}C=C\underset{H}{\overset{R}{<}} \xrightarrow{H_2/Ni\ or\ Pd} \underset{H\quad H}{\overset{R\quad R}{>}}C-C<$$

$$R-C\equiv N \xrightarrow{H_2/Ni\ or\ Pd} R-CH_2NH_2$$

Hydrogen(−I). Hydride ions are among the most powerful reductants known. They are usually used in a complexed form, for example, the tetrahydrido-complex of aluminium:

$$4LiH(s) + AlCl_3(s) \xrightleftharpoons[\text{in ether}]{\text{reflux}} LiAlH_4(\text{ether}) + 3LiCl(s)$$

In the ether solution, tetrahydridoaluminate ions are in equilibrium with a very low concentration of hydride ions.

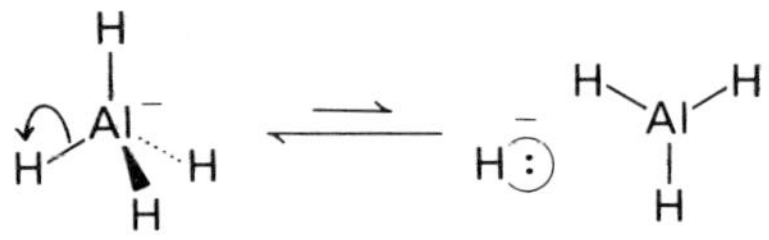

These act as nucleophiles to attack polar unsaturated organic molecules such as ketones, acids and nitriles. For example,

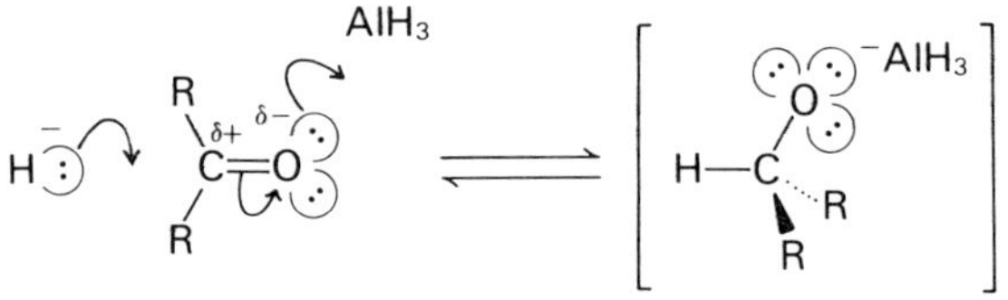

On addition of water, the organo aluminium complexes are hydrolysed. For example, ketones give secondary alcohols (see page 216).

Hydrogen(I). When water, hydrogen halides or acidic solutions react with metals, hydrogen(I) is acting as an oxidant.

$$2\overset{0}{Na}(s) + 2\overset{I}{H_2}O(l) \longrightarrow 2\overset{I}{Na}^+(aq) + 2OH^-(aq) + \overset{0}{H_2}(g)$$

$$\overset{0}{Mg}(s) + \overset{I}{H_2}O(g) \xrightarrow{\text{heat}} \overset{II}{Mg}O(s) + \overset{0}{H_2}(g)$$

$$\overset{0}{Zn}(s) + 2\overset{I}{H}Cl(g) \xrightarrow{\text{heat}} \overset{II}{Zn}Cl_2(s) + \overset{0}{H_2}(g)$$

$$\overset{0}{Fe}(s) + 2\overset{I}{H_3}O^+(aq) \longrightarrow \overset{II}{Fe}^{2+}(aq) + 2H_2O(l) + \overset{0}{H_2}(g)$$

The H(I)/H(0) half-cell is the standard half-cell chosen for measuring redox potentials (see page 93). All other half-cell potentials are compared with that of hydrogen.

10 Industrial Chemistry

10.1 METAL EXTRACTION

Metals are rarely found in an uncombined state in the earth's crust. Only the least reactive occur as a pure metal instead of as an oxide, sulphide, carbonate or sulphate ore. The conversion of a metal ore to the metallic form is known as metal extraction and the process is one of reduction. The reductant chosen for an industrial extraction depends on the nature of the ore, the reactivity of the metal and the economics of the process.

The most reactive metals are extracted using electrolysis because no chemical reductant is powerful enough to achieve extraction under economic conditions of temperature and pressure. For example, sodium, magnesium and aluminium are all extracted electrolytically.

Sodium, the Downs cell

Molten sodium chloride is electrolysed in a circular cell between a central graphite anode and a circular iron cathode. An iron grid separates the anode compartment from the cathode department so that the two products are kept apart. Calcium chloride is added so that the charge melts at a lower temperature. This saves money and also prevents the build up of sodium vapour that would result at an operating temperature of 801°C, the melting-point of sodium chloride.

anode: $2Cl^- \longrightarrow Cl_2 + 2e^-$ $T = 600°C$
cathode: $2Na^+ + 2e^- \longrightarrow 2Na$ power = 7 V × 20 000 A

Magnesium

The major source of magnesium is a carbonate ore. Either this is converted to the chloride and electrolysis is carried out in a cell similar to the Downs cell, or the carbonate is decomposed in a furnace to the oxide. The oxide is then heated with a ferroalloy of silicon at very low pressure and 700°C. Magnesium vapour is distilled from this mixture.

$2MgO(s) + Si(s) \longrightarrow 2Mg(g) + SiO_2(s)$

Aluminium

Bauxite (impure aluminium oxide) is first purified by being treated with concentrated alkali. The aluminate salt that is produced is then seeded with aluminium hydroxide which causes the precipitation of further aluminium hydroxide. Alumina is regenerated by dehydrating the hydroxide.

$$Al_2O_3(s) \text{ impure} \xrightarrow{OH^-} Al(OH)_4^- \xrightarrow{Al(OH)_3} Al(OH)_3(s) \xrightarrow{heat} Al_2O_3(s) \text{ pure}$$

The pure alumina is dissolved in molten cryolite (Na_3AlF_6) and electrolysed in a Hall–Heroult cell. This steel cell has a carbon lining acting as the cathode, and huge carbon anodes which dip into the cell. The anodes are constantly being replaced as they burn in the oxygen generated by the reaction at their surface. Aluminium is produced at the cathode and is siphoned off from the bottom of the cell in a molten, 98% pure, state.

anode: $4AlO_3^{3-} \longrightarrow 2Al_2O_3 + 3O_2 + 12e^-$ $T = 950°C$

$(3C + 3O_2 \longrightarrow 3CO_2)$ power = 5 V × 100 000 A

cathode: $4Al^{3+} + 12e^- \longrightarrow 4Al$

The use of carbon as a reductant

The Ellingham diagram of oxide formation

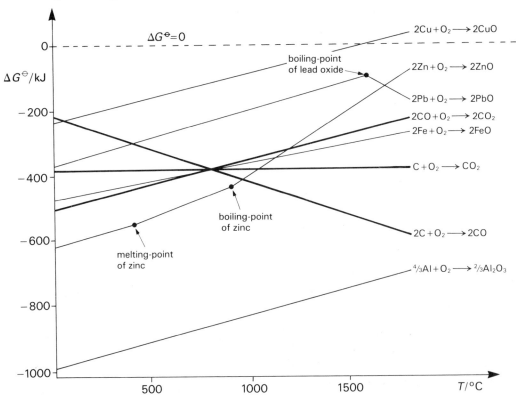

$\Delta G^\ominus$ is the free energy change per mole of each oxide formed at various temperatures, T.

An Ellingham diagram is a graph showing the variation with temperature of the 'free energy' of formation $\Delta G_f^\ominus$ of a number of different oxides. The free energy change of a reaction is related to its enthalpy change $\Delta H^\ominus$ (see page 62) by the equation,

$$\Delta G^\ominus = \Delta H^\ominus - T\Delta S^\ominus$$

$\Delta S^\ominus$ is the change in entropy brought about by the reaction, entropy being a measure of the degree of disorder of the particles within the system. For example, a perfect crystal lattice at a termperature of absolute zero has almost no entropy, whereas a hot gas has an appreciable amount of entropy because of the lack of organization of the particles making it up. Enthalpy and entropy changes, $\Delta H^\ominus$ and $\Delta S^\ominus$, are almost independent of the temperature at which the reaction occurs, but this is not true for the free energy change, $\Delta G^\ominus$. By assuming that $\Delta H^\ominus$ and $\Delta S^\ominus$ are constant over a wide temperature range for a given reaction, differentiation of the above equation with respect to temperature gives the gradient of each line as,

$$\frac{d(\Delta G^\ominus)}{dT} = -\Delta S^\ominus$$

In other words, the slope of a line of the Ellingham diagram is a direct measure of $-\Delta S^\ominus$ for the formation of an oxide: a positive slope indicates a negative entropy change, and a negative slope, a positive entropy change.

The formation of the carbon oxides are emphasized by bold lines and, using these, it is possible to calculate the free energy change for any reaction in which carbon displaces a metal from its oxide. For example, consider the reduction of zinc oxide by carbon to produce zinc and carbon monoxide.

$$ZnO + C \longrightarrow Zn + CO$$

The energetics of the reaction can be treated in terms of the formation of the two oxides at the particular temperature, e.g. at 750°C.

(i) $2Zn + O_2 \longrightarrow 2ZnO$ $\Delta G^\ominus = -480$ kJ (mol of O_2)$^{-1}$
(ii) $2C + O_2 \longrightarrow 2CO$ $\Delta G^\ominus = -380$ kJ (mol of O_2)$^{-1}$

By subtracting (i) from (ii), the required displacement reaction is generated.

$\Longrightarrow$ $2C + 2ZnO \longrightarrow 2Zn + 2CO$ $\Delta G^\ominus = -380 - (-480)$ kJ
$\therefore$ $C + ZnO \longrightarrow Zn + CO$ $\Delta G^\ominus = +50$ kJ mol^{-1}

Since a free energy change indicates whether or not a reaction is likely to occur, this calculation shows whether or not an ore *can* be reduced by carbon at the particular temperature. For example, $\Delta G^\ominus$ is positive for the reduction of zinc oxide at 750°C and so the reaction *cannot* take place.

It is not necessary to perform an elaborate calculation each time the diagram needs to be used. A rule of thumb is easily deduced: if the line for the free energy of formation of a metal oxide is *above* any of the lines for the oxides of carbon, then the reduction of the metal oxide is favourable at that temperature. Three applications of the Ellingham diagram are discussed over the page.

Copper. At about 1650°C, the free energy of formation of copper oxide becomes positive. Above this temperature, therefore, the decomposition of copper oxide to its elements is favourable. In other words, the extraction of copper requires furnance temperatures only.

Iron: the blast furnance. In a blast furnace, iron oxide, carbon, carbon monoxide and carbon dioxide are all present at temperatures that range from about 1400°C at the bottom of the furnace to 500°C at the top. This variation in temperature leads to a variation in the nature of the reactions taking place.

At 500°C Fe/FeO is above CO/CO_2, but below C/CO

$\therefore$ FeO + CO $\longrightarrow$ Fe + CO_2 favourable

but FeO + C $\longrightarrow$ Fe + CO   not favourable

At 1500°C Fe/FeO is below CO/CO_2, but above C/CO

$\therefore$ FeO + CO $\longrightarrow$ Fe + CO_2 not favourable

but FeO + C $\longrightarrow$ Fe + CO   favourable

In other words, as the charge of iron oxide and coke falls through the furnace and carbon monoxide filters up through the charge from the burning coke at the bottom, two processes of reduction take place. At the top, carbon monoxide is the reductant; at the bottom, carbon itself is the reductant.

Lead, zinc and aluminium. Lead and zinc oxides are reduced in blast furnaces similar in design to that used for iron. From the Ellingham diagram, it can be seen that the extraction of lead using carbon monoxide as reductant is favourable at any temperature below about 1800°C. Carbon is also an effective reductant over the whole temperature range. For zinc, a rather different picture emerges. Reduction of the oxide is favourable only at a temperature above the boiling-point of the element. Between 1000 and 1200°C, carbon is the reductant but above 1250°C, it is possible that carbon monoxide also acts as a reductant. Zinc vapour is produced which is drawn off in a stream of carbon monoxide. The vapour is usually shock-cooled by a spray of molten lead from which zinc is separated because molten zinc and lead are almost immiscible.

Aluminium can only be produced in a blast furnace at temperatures in excess of 2000°C. Not only is electrolysis a far cheaper method of reduction than maintaining a furnace at 2000°C, but also aluminium carbide would form under such extreme conditions.

10.2 ALKALIS

There are three important alkalis that are manufactured on a large scale: sodium hydroxide, ammonia and sodium carbonate. Sodium hydroxide is manufactured in conjunction with hydrogen and chlorine from brine; sodium carbonate is also a product of the salt industry (it is made by the Solvay process), and ammonia is synthesized by the Haber process.

The diaphragm cell

Concentrated brine is electrolysed in a cell whose electrodes are separated by a porous asbestos diaphragm. The brine flows into the top part which contains the anode, and about half the chloride ions are discharged. The chlorine is immediately piped out of the top of the cell. Under the aqueous conditions, the discharge of hydrogen at the cathode is more favourable than that of sodium (see page 107). The hydrogen is trapped by the diaphragm and led off separately, while the cell liquor drained from the bottom of the cell contains about 10% of sodium hydroxide and 15% of sodium chloride. Sodium chloride is less soluble than sodium hydroxide and is removed by fractional crystallization.

anode: $2Cl^-(aq) \longrightarrow Cl_2(g) + 2e^-$ $15°C < T < 90°C$

cathode: $2H_2O(l) + 2e^- \longrightarrow 2OH^-(aq) + H_2(g)$ power $= 4\,V \times 150\,000\,A$

The Solvay process

A solution of brine is saturated with ammonia and then pumped into the top of a large tower whose inside contains an extensive system of baffles. Carbon dioxide is pumped into the bottom of the tower and the stream of falling liquid meets the gas passing up through the column. The carbon dioxide dissolves and produces an acidic solution containing hydrogencarbonate ions. The ammonia present in the brine ensures that the position of equilibrium (shown below) lies well to the right,

$$CO_2(g) + 2H_2O(l) \rightleftharpoons HCO_3^-(aq) + H_3O^+(aq)$$

because the ammonia molecules react rapidly with the hydroxonium ions.

$$NH_3(aq) + H_3O^+(aq) \rightleftharpoons NH_4^+(aq) + H_2O(l)$$

The bottom section of the tower is cooled by water vanes so that sodium hydrogencarbonate precipitates as a result of its low solubility.

$$Na^+(aq) + HCO_3^-(aq) \longrightarrow NaHCO_3(s)$$

A suspension of sodium hydrogencarbonate is drained off and filtered. Sodium carbonate is produced by thermal decomposition.

$$2NaHCO_3(s) \xrightarrow{heat} Na_2CO_3(s) + CO_2(g) + H_2O(g)$$

The Haber process

Nitrogen and hydrogen are mixed in a ratio of 1:3 and passed over a finely divided iron catalyst at a temperature of ~500°C and a pressure between 20 and 100 MPa. The gases are rapidly cooled and ammonia liquefies, while the unreacted nitrogen and hydrogen are recycled. The raw materials are coke or hydrocarbons, air and water: nitrogen is obtained by fractional distillation of the air, and hydrogen is extracted from water by reduction using coke or hydrocarbons.

$$N_2(g) + 3H_2(g) \underset{}{\overset{iron}{\rightleftharpoons}} 2NH_3(g) \quad \begin{array}{l} T \approx 500°C \\ p \approx 35\,MPa \end{array}$$

The forward reaction is exothermic, so a high temperature leads to a low yield of ammonia (see page 83). However, to run the plant under the most economic conditions, it is necessary to balance the effects of high temperature against the length of time needed for the reactants to reach equilibrium at the catalytic surface. A higher temperature leads to a faster rate of reaction and, therefore, the reactants need less time in the reaction vessel to reach equilibrium. However, at high temperatures, the position of equilibrium lies to the left and so more cycles are needed to achieve complete conversion. The optimum temperature for maximum efficiency is about 500°C.

10.3 ACIDS

Hydrochloric acid

A jet of hydrogen is burnt in chlorine to produce hydrogen chloride gas which is then dissolved in water. Since both hydrogen and chlorine are products of the electrolysis of brine (page 169), this one process is responsible for the manufacture of hydrogen, chlorine, sodium hydroxide and hydrochloric acid.

Sulphuric acid, the Contact process

The raw materials used in the manufacture of sulphuric acid are sulphur dioxide, air and water. Sulphur dioxide is produced by burning sulphur (extracted by the Frasch process from sulphur deposits, or obtained from the hydrogen sulphide found in natural gas, or as a by-product when refining petroleum). Air and sulphur dioxide are purified, dried and then passed over beds of vanadium(v) oxide catalyst in four separate exposures. After each one, the mixture is led out and cooled so that the temperature of the catalyst does not exceed 450°C. Sulphur trioxide is produced by an exothermic reaction and the cooling is necessary for two reasons. Firstly, the catalyst decomposes above 500°C and secondly, like the ammonia synthesis (described above), a high temperature favours the reactants.

$$SO_2(g) + \tfrac{1}{2}O_2(g) \underset{}{\overset{V_2O_5}{\rightleftharpoons}} SO_3(g) \quad \Delta H^\ominus = -98 \text{ kJ mol}^{-1}$$

The reactants and reaction chamber must be thoroughly dry because sulphur trioxide reacts violently with water to give an acidic mist of sulphuric acid droplets. In order to avoid this happening when the trioxide is finally converted into sulphuric acid during the last stage of the process, a supply of concentrated acid is used as a solvent for the trioxide. Water is then added in appropriate quantity to complete the reaction.

$$SO_3(g) + H_2SO_4(l) \longrightarrow \underset{\text{oleum}}{H_2S_2O_7(l)} \xrightarrow{H_2O} 2H_2SO_4(l)$$

The unreacted gas is recycled, while the sulphuric acid produced is cooled before being stored or pumped back into the tower as solvent.

Nitric acid

Nitric acid is manufactured by a process that has some similarity to that by which sulphuric acid is made; the dioxide of the non-metal is combined with air and water, and the unreacted gases are recycled. The raw materials are, in fact, air and ammonia which is oxidized in the presence of a platinum-rhodium catalyst to nitrogen monoxide and steam. The catalyst is heated electrically to start with, but the exothermic oxidation makes further heating unnecessary.

$$4NH_3(g) + 5O_2(g) \xrightarrow{Pt-Rh} 4NO(g) + 6H_2O(g) \qquad T = 850°C \\ p = 0.1 \text{ MPa}$$

In the presence of more air and water, the nitrogen monoxide is converted first to the dioxide and then to nitric acid.

$$2NO(g) + O_2(g) \longrightarrow 2NO_2(g)$$

$$4NO_2(g) + O_2(g) + 2H_2O(l) \rightleftharpoons 4HNO_3(l)$$

The nitric acid that is produced has a high concentration of dissolved nitrogen dioxide. This is removed by distillation leaving an azeotropic solution of nitric acid (68%) and water.

10.4 PETROCHEMICALS

The term, petrochemical has now become associated with any product or by-product derived from a fossil fuel source: coal, oil or natural gas. The fact that these three naturally occurring substances are known as fossil fuels indicates one of their major uses. Oil, for example, is fractionally distilled to provide gaseous fuel (C_1–C_4), petrol for

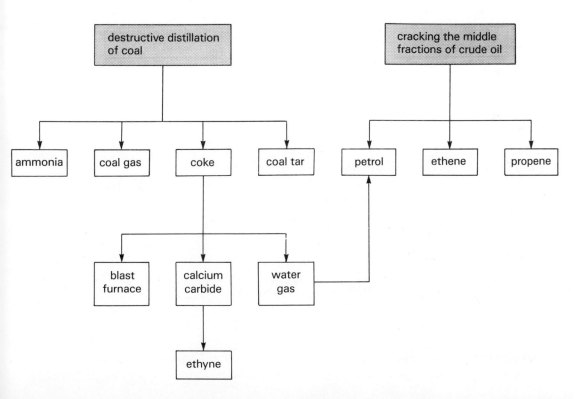

cars (C_4–C_{12}), paraffin and jet fuel (C_{11}–C_{15}), fuel for central heating burners (C_{15}–C_{20}) and fuel for power stations and ships (C_{20}–C_{40}). A variety of names is given to these different boiling fractions. For example, the following names are common for the different fuels mentioned above: Calor gas, gasoline, kerosene, diesel and fuel oil. Coal and natural gas are both used as fuels in their own right as well. However, coal and the middle fractions of oil are also decomposed under various conditions to produce a range of raw materials that provide the starting-points for the petrochemical industry. The decomposition process is usually known as destructive distillation when referred to coal, and cracking when referred to oil fractions. The major products are shown on page 171.

Derivatives of ethene

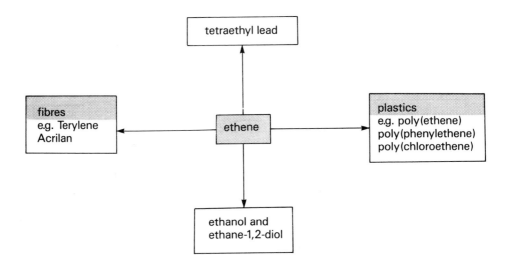

Fibres
Terylene:

Acrilan:

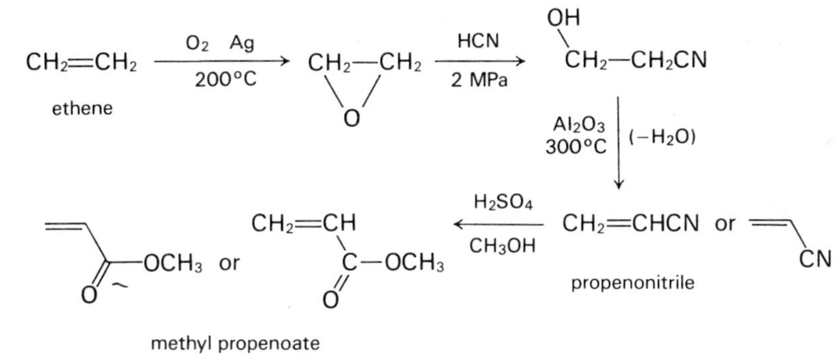

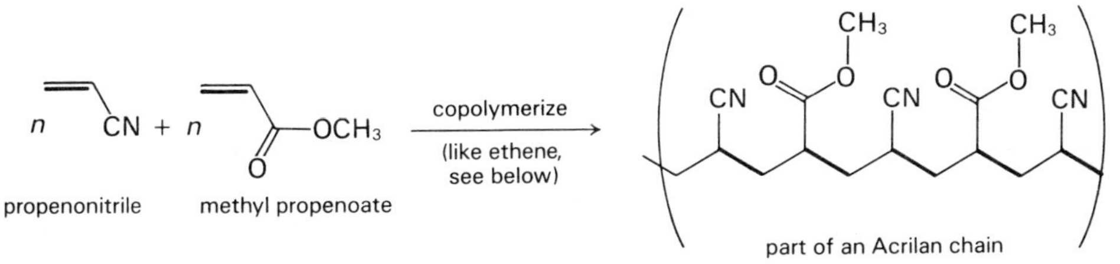

$$CH_2{=}CH_2 \xrightarrow[200°C]{O_2\ Ag} \underset{\text{(epoxide)}}{CH_2{-}CH_2} \xrightarrow[2\ MPa]{HCN} \underset{}{CH_2{-}CH_2CN}\ (OH)$$

$$\downarrow Al_2O_3,\ 300°C\ (-H_2O)$$

methyl propenoate: $CH_2{=}CH{-}C({=}O){-}OCH_3$ ← $\xrightarrow[CH_3OH]{H_2SO_4}$ $CH_2{=}CHCN$ (propenonitrile)

$$n\ \text{(propenonitrile)} + n\ \text{(methyl propenoate)} \xrightarrow[\text{(like ethene, see below)}]{\text{copolymerize}} \text{part of an Acrilan chain}$$

Plastics. Poly(ethene), poly(chloroethene) and poly(phenylethene) are three common plastics derived from ethene. The unsaturated compounds ethene, chloroethene or phenylethene are polymerized using a Ziegler–Natta catalyst (see page 193) at a pressure of between 0.5 and 2.0 MPa. For example,

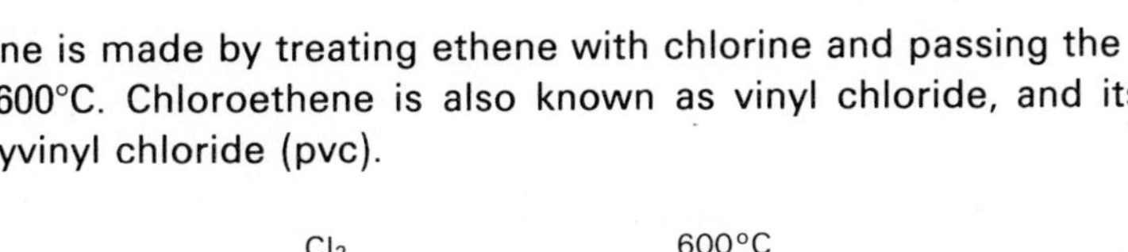

$$n\ CH_2{=}CHCl \xrightarrow[TiCl_3]{Al(C_2H_5)_3} \text{poly(chloroethene) or polyvinylchloride (pvc)}$$

chloroethene ('vinyl chloride')

Chloroethene is made by treating ethene with chlorine and passing the product over a cracker at 600°C. Chloroethene is also known as vinyl chloride, and its polymer is, therefore, polyvinyl chloride (pvc).

$$CH_2{=}CH_2 \xrightarrow{Cl_2} CH_2Cl{-}CH_2Cl \xrightarrow[(-HCl)]{600°C} CH_2{=}CHCl$$

ethene → → chloroethene ('vinyl' chloride)

Phenylethene is made by combining ethene with benzene, and then dehydrogenating the addition product. Phenylethene is also known as styrene, and its polymer is sometimes called polystyrene.

$$CH_2\!=\!CH_2 \text{ (ethene)}$$

$$\text{benzene} \xrightarrow{AlCl_3} \text{ethylbenzene (} C_6H_5CH_2CH_3\text{)} \xrightarrow[600°C\ (-H_2)]{ZnO} \text{phenylethene ('styrene') (} C_6H_5CH\!=\!CH_2\text{)}$$

Ethanol, ethane-1,2-diol and tetraethyl lead. These substances find use as a solvent, as an antifreeze and as a petrol anti-knock respectively. Ethanol and glycol are also important industrial raw materials as well as solvents.

$$CH_2\!=\!CH_2 \xrightarrow[300°C\quad 7\ MPa]{H_2O\quad H_3PO_4} CH_3\text{—}CH_2OH \text{ (ethanol)}$$

$$CH_2\!=\!CH_2 \xrightarrow[200°C]{O_2\ Ag} \underset{O}{CH_2\text{—}CH_2} \xrightarrow[2\ MPa]{H_2O\ 200°C} CH_2OH\text{—}CH_2OH \text{ (ethane-1,2-diol)}$$

$$CH_2\!=\!CH_2 \xrightarrow{HCl} CH_3\text{—}CH_2Cl \xrightarrow[(-NaCl)]{Pb\text{-}Na\ alloy} Pb(C_2H_5)_4 \text{ (tetraethyl lead)}$$

Propene

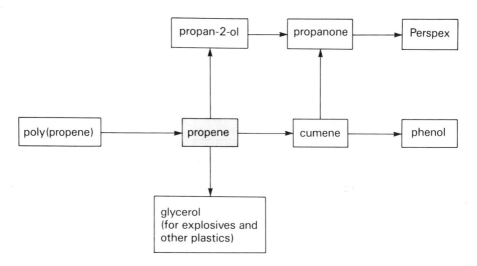

Poly(propene) and other propyl plastics are made in the same way as those derived from ethene. However, the two major outlets for propene are in the *cumene-phenol* process (whose products are phenol and propanone), and in the different oxidation products that can be obtained.

Oxidation. Propene is oxidized by chlorine, firstly under homolytic conditions and then subsequently, under alkaline conditions, to make 'glycerol' (propane-1,2,3-triol).

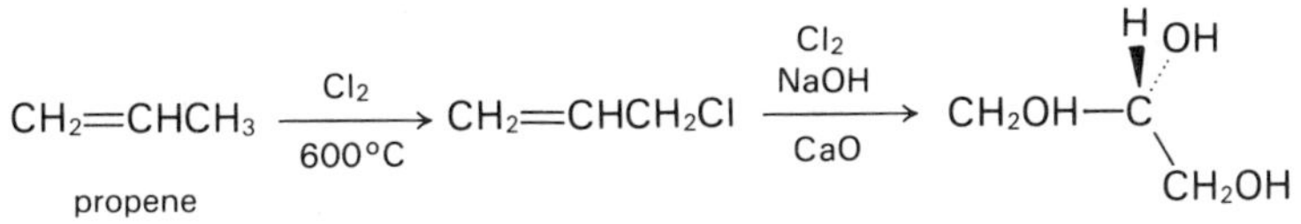

$$CH_2{=}CHCH_3 \xrightarrow[600°C]{Cl_2} CH_2{=}CHCH_2Cl \xrightarrow[CaO]{Cl_2, NaOH} CH_2OH{-}\overset{H}{\underset{CH_2OH}{C}}{-}OH$$

propene   propane-1,2,3-triol ('glycerol')

Glycerol is used to manufacture a number of explosives, plastics and soaps. Alternatively, propene is oxidized to propanone (acetone) via propan-2-ol.

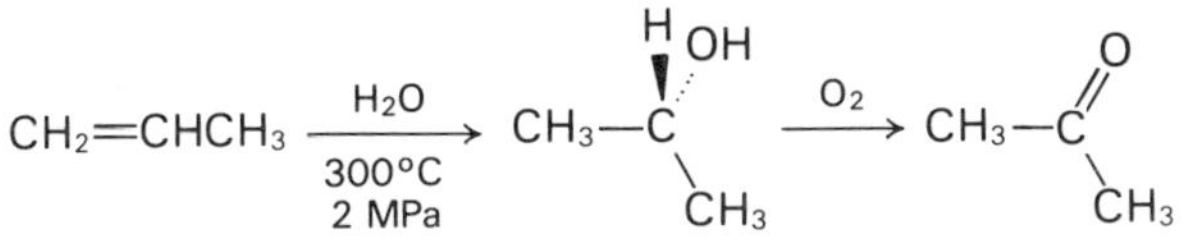

$$CH_2{=}CHCH_3 \xrightarrow[\substack{300°C \\ 2\ MPa}]{H_2O} CH_3{-}\overset{H}{\underset{CH_3}{C}}{-}OH \xrightarrow{O_2} CH_3{-}\overset{O}{\underset{CH_3}{C}}$$

propanone ('acetone')

Both propan-2-ol and propanone are used as solvents and as raw materials for other processes. One process of particular value is the manufacture of Perspex from propanone.

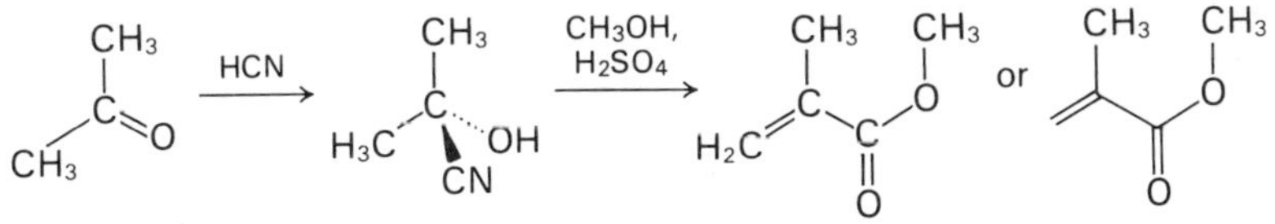

propanone 'acetone' → (HCN) → H₃C–C(CH₃)(OH)(CN) → (CH₃OH, H₂SO₄) → methyl 'methacrylate'

$$n \left(\begin{array}{c} CH_3\ CH_3 \\ \diagup\diagdown \\ O \\ \| \\ O \end{array} \right) \xrightarrow{polymerize} \text{part of a Perspex chain}$$

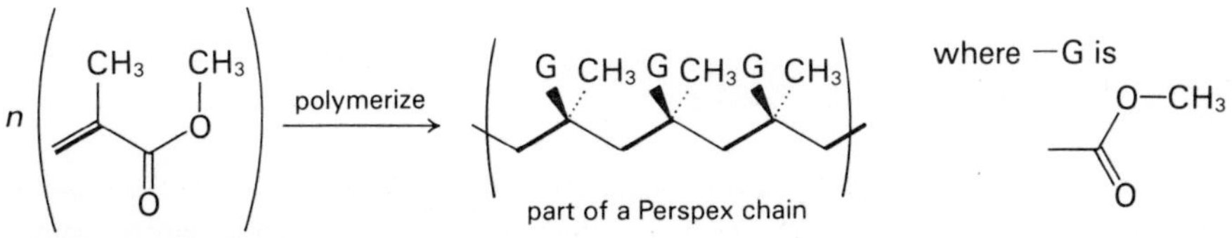

methyl 'methacrylate'

where —G is —O—CH₃ (ester group)

The cumene-phenol process. Phenol is a very important raw material for the pharmaceutical industry as well as for the production of dyes and plastics. Benzene is converted to phenol via 'cumene' [(methylethyl)benzene] which is produced by reacting propene with benzene.

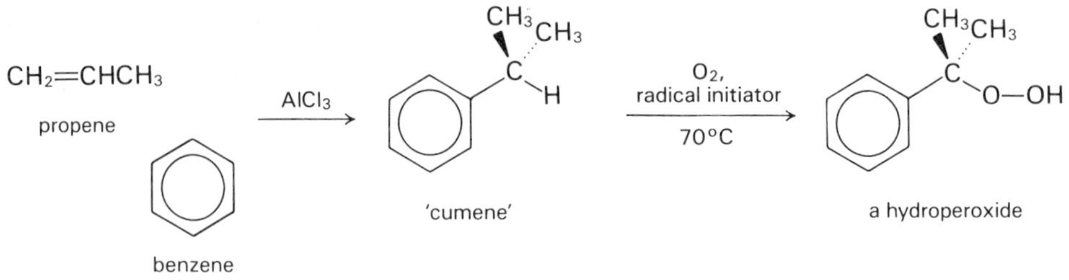

[Reaction scheme: propene CH$_2$=CHCH$_3$ + benzene →(AlCl$_3$) cumene →(O$_2$, radical initiator, 70°C) a hydroperoxide]

The unstable hydroperoxide is hydrolysed under acidic conditions.

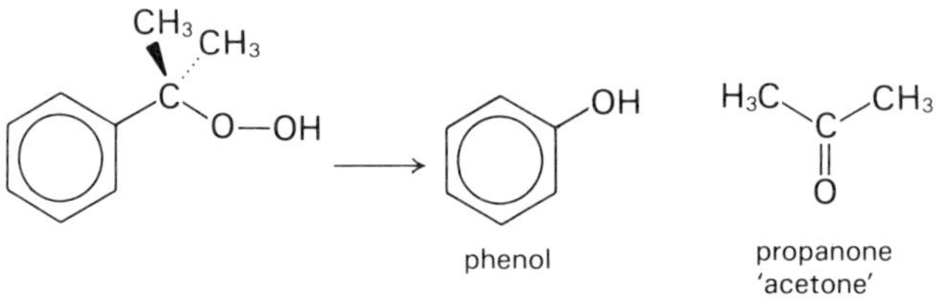

[Reaction scheme: cumene hydroperoxide → phenol + propanone 'acetone']

Coal derivatives

Coal tar, water gas and the ethyne derived from calcium carbide are all used as a source of petrochemicals.

Coal tar is a complex mixture largely of aromatic compounds. The lower boiling fractions provide benzene, methylbenzene (toluene) and dimethylbenzene (xylene), which are used as solvents and in the manufacture of phenols and detergents. The higher boiling fractions provide naphthalene and anthracene. These are required for the synthesis of a number of plastics and dyes.

Water gas derives its name from the way in which it is produced. Steam is passed over white-hot coke, or is mixed with long-chain hydrocarbons in a special cracker.

$$C(s) + H_2O(g) \xrightarrow{1000°C} CO(g) + H_2(g)$$

$$[C_{30-40}H_{20-60}] \xrightarrow[1000°C]{H_2O(g)} CO(g) + H_2(g)$$

Increasing amounts of water gas are now being converted into petrol via the Fischer–Tropsch synthesis. The price of crude oil is becoming so high that petrol synthesized from water gas may soon be a viable alternative.

$$nCO(g) + (2n+1)H_2(g) \xrightarrow[\substack{80°C \\ \text{high} \\ \text{pressures}}]{Co-Ni} C_nH_{2n+2}(g) + nH_2O(g)$$

10 Industrial Chemistry

Ethyne is almost as important a raw material as ethene. It is converted into plastics such as Perspex and pvc, and into fibres such as Orlon, Acrilan and various nylons. The conversion of ethyne into 'chloroprene' is the first step in the manufacture of a widely-used artificial rubber known as Neoprene.

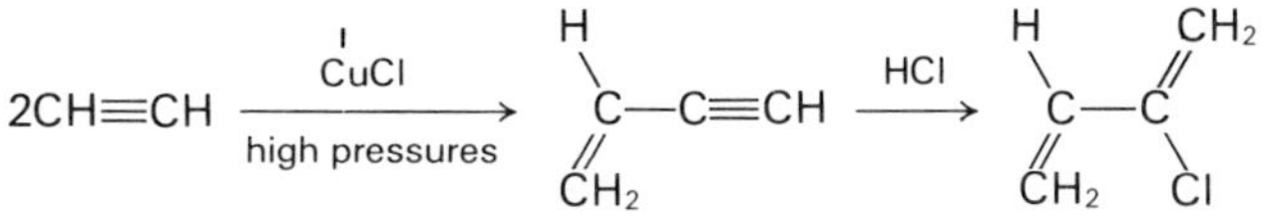

$$2CH\equiv CH \xrightarrow[\text{high pressures}]{CuCl} \underset{CH_2}{\overset{H}{C}}=C-C\equiv CH \xrightarrow{HCl} \underset{CH_2}{\overset{H}{C}}=\underset{Cl}{\overset{CH_2}{C}}$$

'chloroprene'

Chloroprene is polymerized to give Neoprene.

A typical nylon is called nylon-66 and it is made from ethyne as follows.

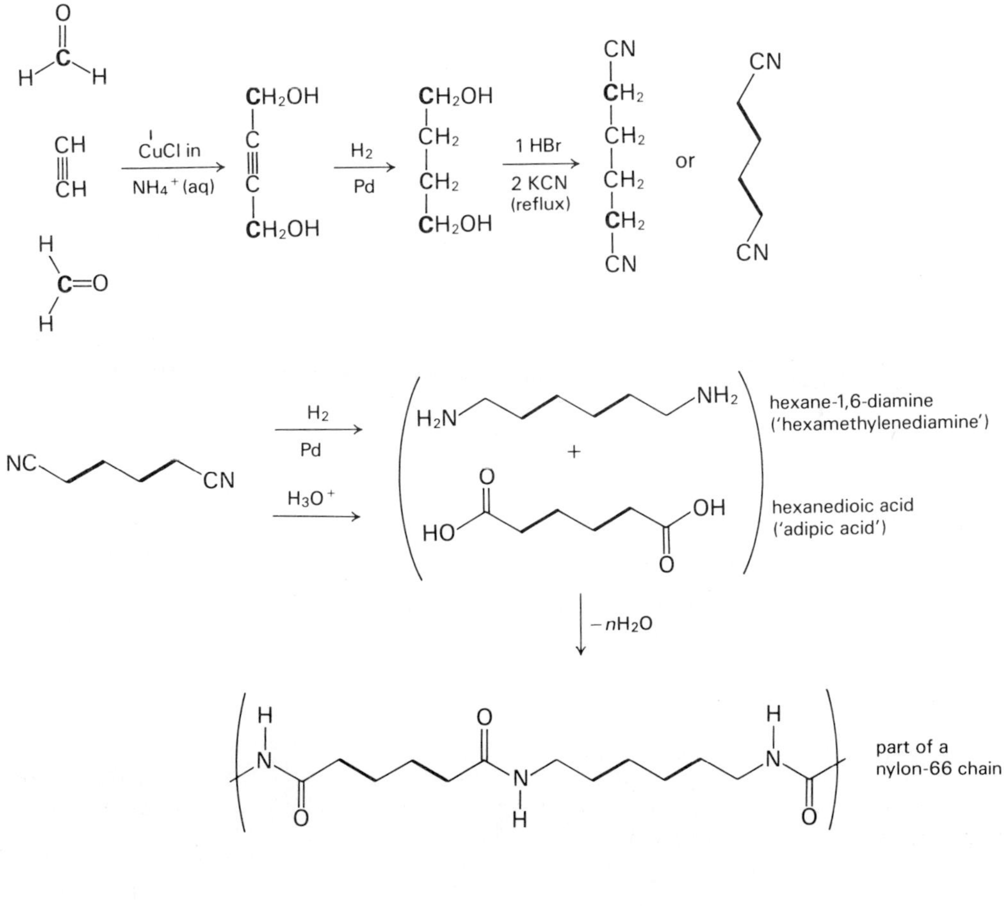

[reaction scheme: HCHO + CH≡CH → (via CuCl in NH₄⁺(aq)) HOCH₂–C≡C–CH₂OH → (H₂/Pd) HOCH₂CH₂CH₂CH₂OH → (1 HBr, 2 KCN reflux) NC–CH₂CH₂CH₂CH₂–CN or equivalent dinitrile]

$$NC\text{–}(CH_2)_4\text{–}CN \xrightarrow[H_3O^+]{H_2 \; Pd} \left(\begin{array}{c} H_2N\text{–}(CH_2)_6\text{–}NH_2 \\ + \\ HOOC\text{–}(CH_2)_4\text{–}COOH \end{array} \right)$$

hexane-1,6-diamine ('hexamethylenediamine')

hexanedioic acid ('adipic acid')

↓ −nH₂O

[–NH–(CH₂)₆–NH–CO–(CH₂)₄–CO–]ₙ part of a nylon-66 chain

11 Reactivity of Organic Compounds

11.1 MECHANISM

A reaction mechanism explains how reactants become products by illustrating the movement of electron density occurring during a collision between reactant particles. A mechanism suggests which bonds break, which bonds form and in what order the events take place.

There are two major types of mechanism: homolytic and heterolytic. Heterolytic mechanisms can be divided further into nucleophilic and electrophilic mechanisms.

Homolytic mechanisms

During the homolytic fission of a bond, the two fragments that are produced each receive one electron from the bonding pair. For example, chlorine in sunlight, or ethanoyl peroxide at 150°C (see page 193), undergo homolytic fission.

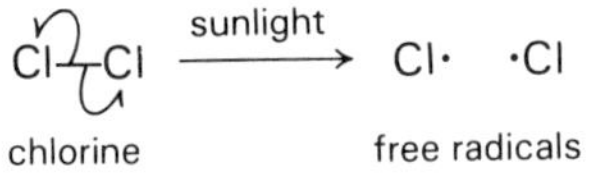

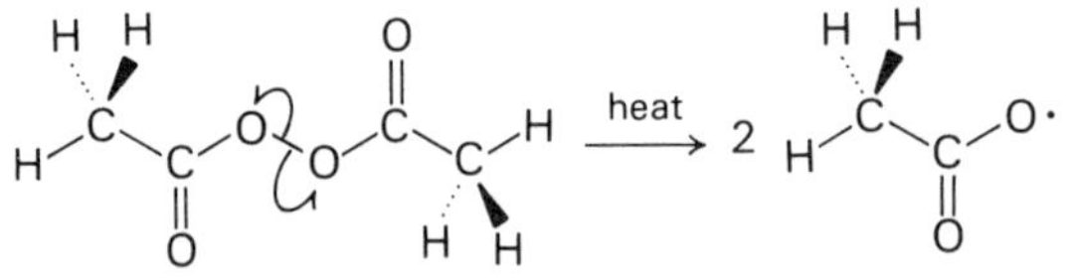

Heterolytic fission leads to one atom receiving both bonding electrons. For example, hydrogen chloride in water:

A mechanism that proceeds via homolytic fission of bonds is called a free radical or homolytic mechanism. The example of methane and chlorine is described on page 79.

Heterolytic mechanism

The mechanisms of the vast majority of organic reactions are heterolytic in nature. A pair of electrons on one particle is attracted to an electron-deficient site on another particle. Electron pairs may then be transferred as a result of the attraction. The collision between the reactant particles must satisfy both the energy and orientation factors specific to the reaction (see the discussion of collision theory on page 73). In drawing a mechanism, therefore, it is usual to show three distinct stages: reactants, an activated complex, and products.

The reactant particles. These are drawn in such a way that the orientational requirements are satisfied: the electron pair of the one particle is lined up with the electron-deficient site of the other. For example, in the collision of a hydroxide ion with a bromoethane molecule, the following is drawn:

The remainder of the reactant particles is shown in shorthand so that electron pairs and electron-deficiency are emphasized, and the diagram is not cluttered by unnecessary detail.

A curly arrow is used to show the movement of a pair of electrons. The tail of the arrow is drawn at the original position of the electron pair and the head is drawn at their destination, for example:

The activated complex. Unless the formation of a new bond and fission of an existing bond are simultaneous, a high energy particle containing partially formed bonds or localized charge distribution is formed. This is an activated complex (see page 76) and is drawn after the reactant particles. An activated complex is shown in square brackets, for example:

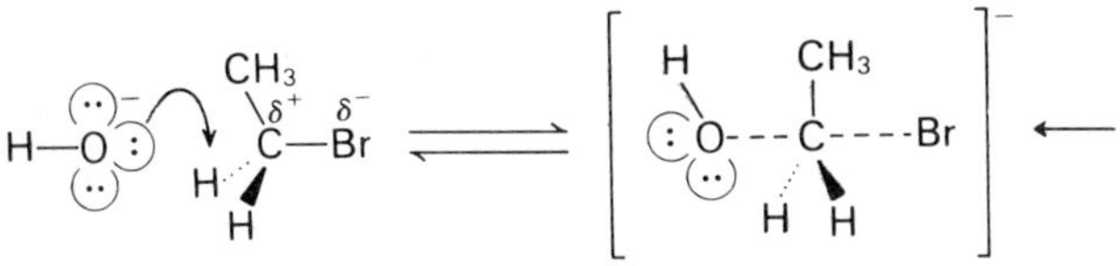

[Diagram: reaction mechanism showing hydroxide attacking CH₃CH₂Br forming activated complex]

The break-up of the activated complex (to form products) is also shown using curly arrows.

The product particles. These are drawn as the third stage of the whole mechanism. It is usual to retain the orientation of the particles in this final drawing.

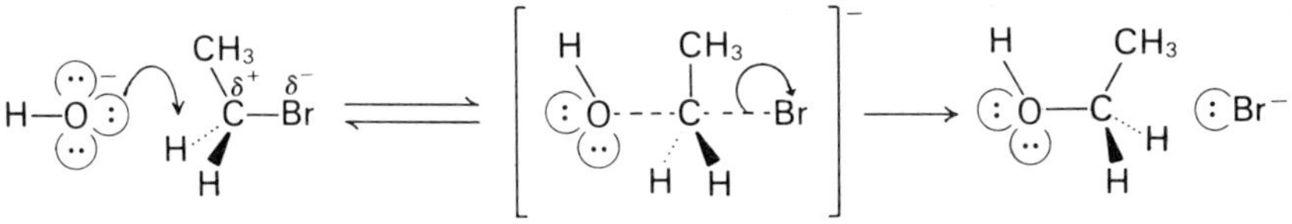

[Diagram: full mechanism showing hydroxide + CH₃CH₂Br → activated complex → CH₃CH₂OH + Br⁻]

11.2 NUCLEOPHILIC REACTIONS

Nucleophiles and electron-deficient sites

A nucleophile is a particle with a lone pair of electrons which become σ-electrons after a collision with the electron-deficient site of another particle. Typical nucleophiles include those shown below. Note that it is often necessary to show a localized charge on the nucleophilic atoms after their electron pair donation. Two bonded atoms have control (on average) of one bonding electron each; therefore, when one atom donates both electrons to form a new bond, it loses effective control of one electron.

nucleophile	electron-deficient site	activated complex
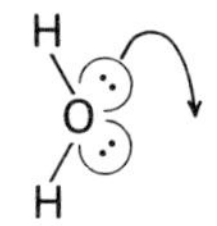 H₂O (with lone pairs shown)	E^+	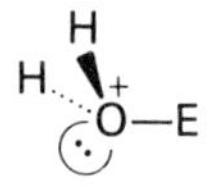 H₂O⁺—E
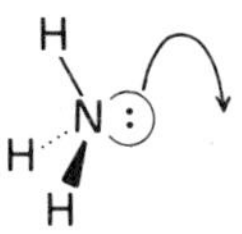 NH₃ (with lone pair shown)	E^+	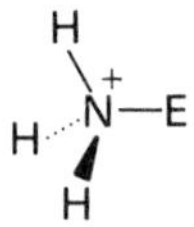 H₃N⁺—E
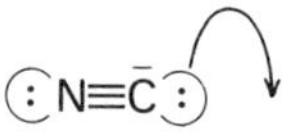 :N≡C:	E^+	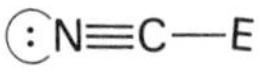 :N≡C—E

11 Reactivity of Organic Compounds

Electron-deficiency is caused either by an inductive effect (I) or by a mesomeric effect (M).

An inductive effect occurs when a bond is polarized because one atom tends to take a greater share of the bonding σ-electrons. Three typical cases are outlined below.

1 When two atoms of different electronegativity bond together.

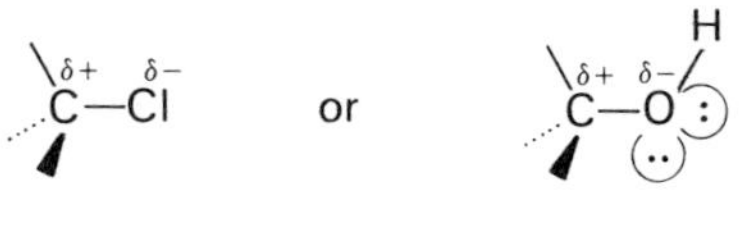

2 When an atom can form a stable leaving group by breaking free with both the bonding electrons.

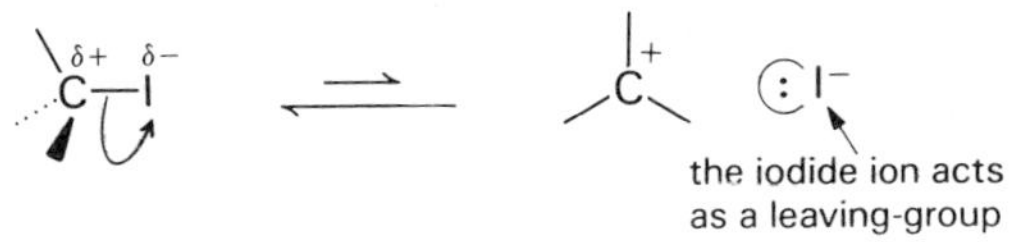

the iodide ion acts as a leaving-group

3 When an unequal charge distribution results from the approach of a charged particle.

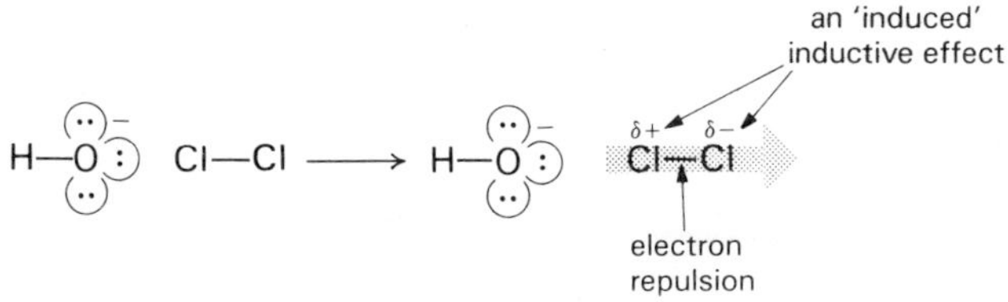

an 'induced' inductive effect

electron repulsion

The symbols $+I$ and $-I$ are used to denote the direction of an inductive effect. An atom that exerts a $+I$ effect on another releases σ-electron density towards it. For example, in the chlorine molecule above, the first chlorine atom has an induced $+I$ effect on the second chlorine atom. Conversely, in the iodoalkane molecule, the iodine atom exerts a $-I$ effect on the carbon atom because it tends to remove σ-electron density from the carbon atom.

Mesomeric effects concern the movement of π-electrons in a molecule. The two symbols $+M$ and $-M$ are used in the same way that $+I$ and $-I$ are used to show the direction of inductive effects. For example, the oxygen atom in a ketone molecule exerts a $-M$ effect on the carbon atom. This supplements the $-I$ effect that the oxygen atom is already exerting.

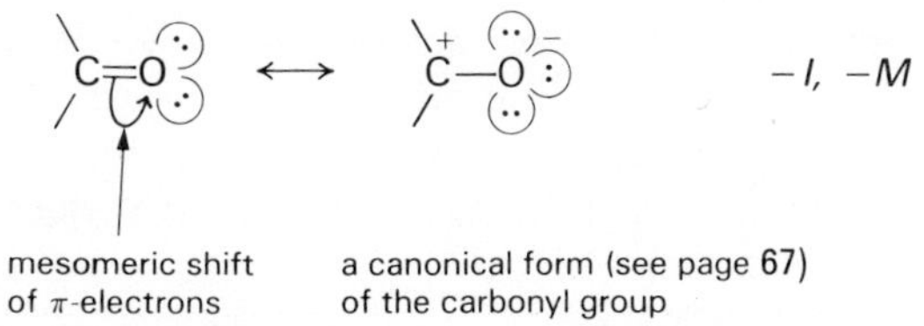

$-I, -M$

mesomeric shift of π-electrons

a canonical form (see page 67) of the carbonyl group

However, in an *enol*-molecule, the oxygen atom exerts a +M effect by donating a lone pair into the system. In this case, the oxygen atom exerts opposing effects: −I and +M. The mesomeric effect is usually dominant (for example, as in phenols on page 201).

$$-I, +M$$

Nucleophilic substitution

A saturated electron-deficient carbon atom is open to attack by a nucleophile with the result that substitution takes place. The electron-deficiency of a saturated carbon atom can only be caused by an inductive effect. There are two possible mechanisms by which substitution can occur (Nu stands for a nucleophile; Le stands for a leaving group):

In the first mechanism, the leaving group leaves after a collision between a nucleophile and the reactant molecule. The order of this reaction is two, because the rate depends on the concentration of both the nucleophile and the reactant. In the second mechanism, however, it is suggested that the leaving group actually has a tendency to leave of its own accord. The carbonium ion produced would then react very quickly indeed with any nucleophile close by. The order of this reaction is only one, because the first step is rate-determining (see page 75), being so much slower than the second step. The overall rate depends only on the concentration of the reactant and is independent of the concentration of the nucleophile.

For these reasons, the two possible mechanisms are known as S_N2 and S_N1 respectively. S_N stands for nucleophilic substitution, and the figure gives the order of the reaction. For a particular nucleophilic substitution, an experiment can be designed to measure the order of the reaction, and this will then indicate which of the two possible mechanisms is dominant.

Nucleophilic addition

An unsaturated, electron-deficient carbon atom is even more open to nucleophilic attack. In order to accommodate an attacking nucleophile, it is only necessary for a mesomeric shift of π-electrons to take place. In many cases, the resulting activated complex is rapidly protonated and this leads, in effect, to the addition of H—Nu. For example, the carbonyl and nitrile group are susceptible to nucleophilic addition:

addition products

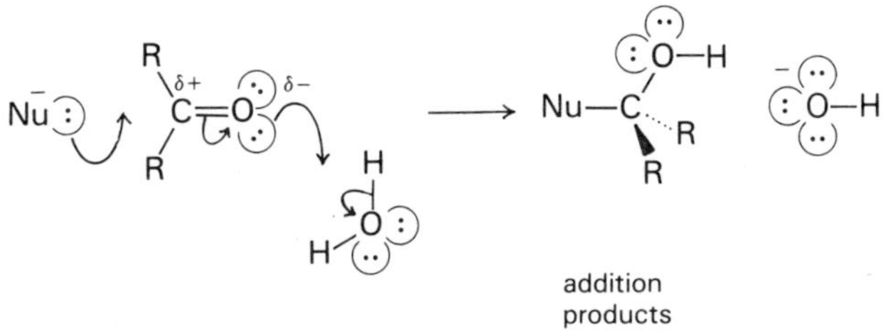

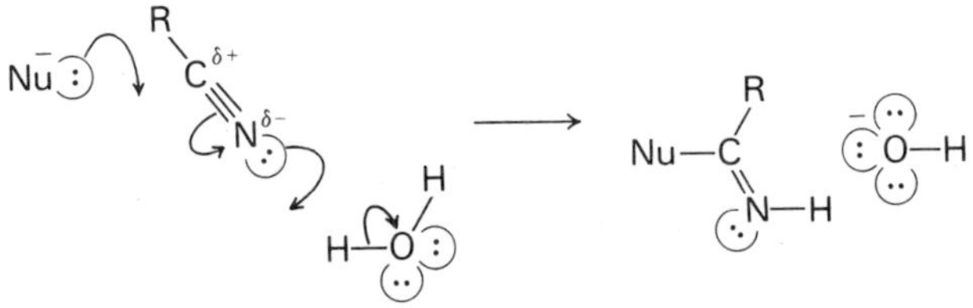

In many reactions that involve nucleophilic addition, further rearrangement occurs. Sometimes, this takes the form of elimination, as in the case of the reaction of hydroxylamine with a ketone on page 213: the addition product eliminates water to produce an oxime as shown below.

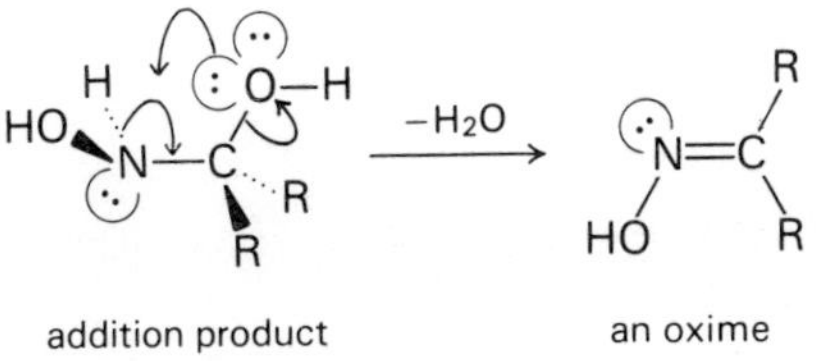

addition product   an oxime

Sometimes, internal proton transfer occurs, as in the case of the reaction of hydroxide ions with a nitrile (page 227):

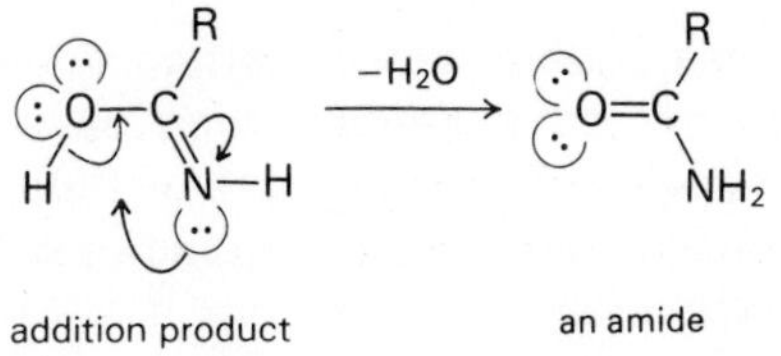

addition product   an amide

11.3 ELECTROPHILIC REACTIONS

Electrophiles and π-electrons

An electrophile is an electron-deficient particle that can accept a pair of π-electrons from another particle. Although many particles are electron-deficient (for example, metal cations), the criterion for electrophilic behaviour is their ability to accept a π-pair in the formation of a new σ-bond. The most common electrophiles are listed below.

1 strong acids

$$\overset{\delta+}{H}-\overset{\delta-}{A}$$

2 halogen molecules act as 'induced' electrophiles (compare with the chlorine molecule on page 181)

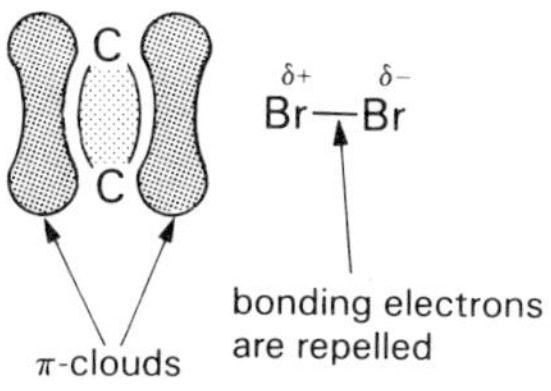

π-clouds

bonding electrons are repelled

3 chloroderivatives in the presence of a Lewis acid (an electron-pair acceptor) such as aluminium chloride

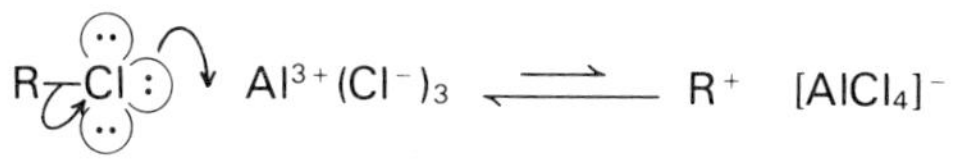

$$R-Cl: \quad Al^{3+}(Cl^-)_3 \rightleftharpoons R^+ \quad [AlCl_4]^-$$

4 nitryl (nitronium) ions present in the mixture of concentrated nitric and sulphuric acids

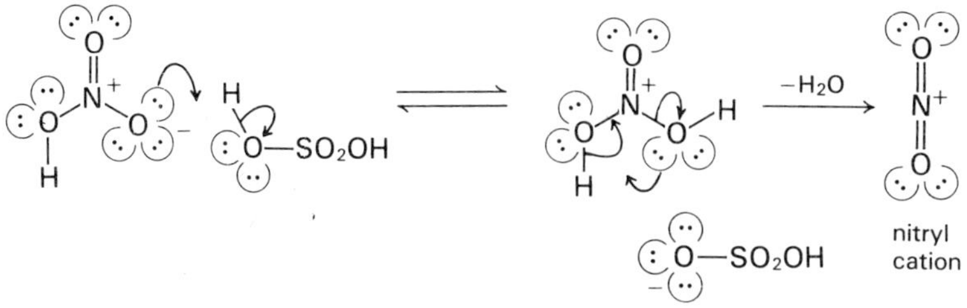

nitryl cation

The π-electrons of an unsaturated carbon–carbon bond are quite readily attracted towards an electrophile. Those of unsaturated carbon–oxygen or carbon–nitrogen bonds are more difficult to draw away from the influence of the more electronegative oxygen and nitrogen atoms. The initial movement of π-electrons leads to the formation of a π-complex in which the distribution of π-electrons is shown by a 'pecked' triangle:

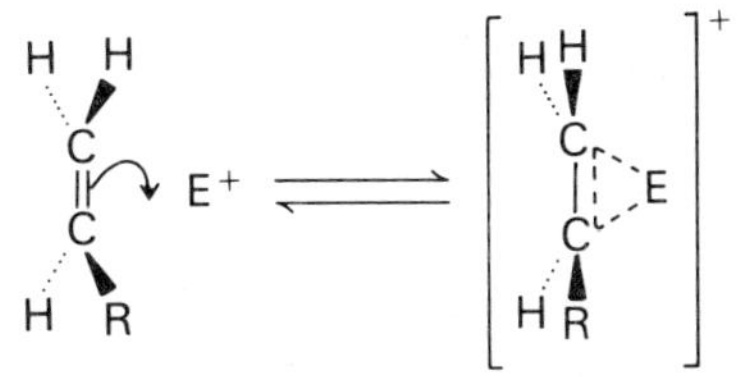

In the case of an aromatic π-system (see page 194), the π-complex is less easy to represent:

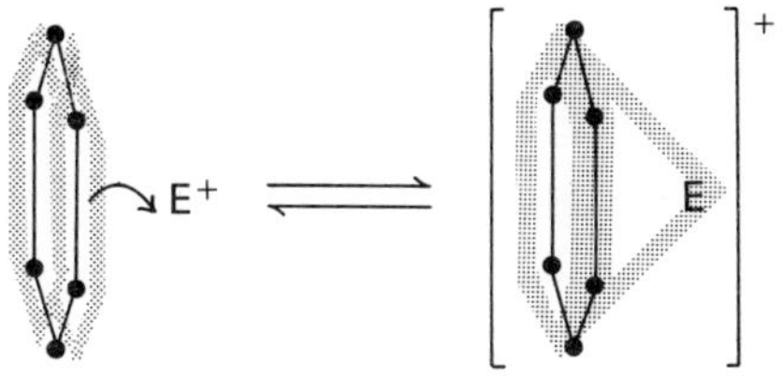

Both these π-complexes tend to rearrange to form σ-complexes:

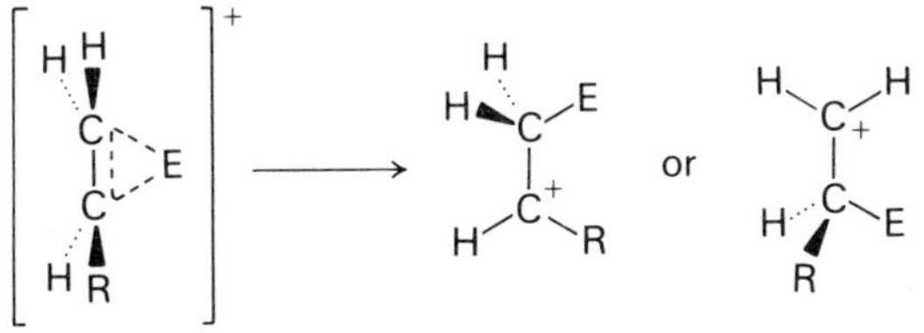

π-complex   σ-complexes

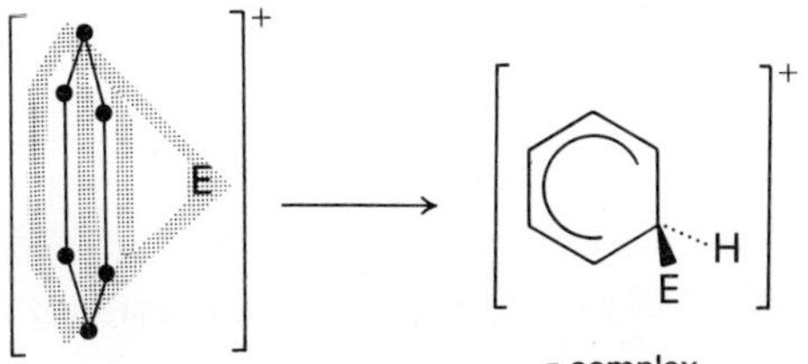

π-complex   σ-complex ('Wheland' intermediate)

The name 'σ-complex' is used because the electrophile is bonded to the reactant particle by a σ-bond instead of by a delocalized π-cloud.

Electrophilic addition

Of all compounds, alkenes show the most marked tendency to undergo electrophilic addition. Typically, the electrophiles involved are strong acids or halogens. For example, the reactions of an alkene with concentrated sulphuric acid and with bromine are shown overleaf.

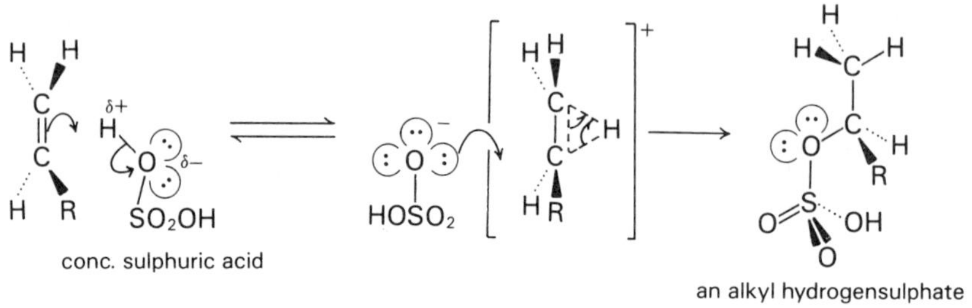

[Reaction schemes showing addition of conc. sulphuric acid to an alkene giving an alkyl hydrogensulphate, and addition of bromine to an alkene giving a dibromoalkane.]

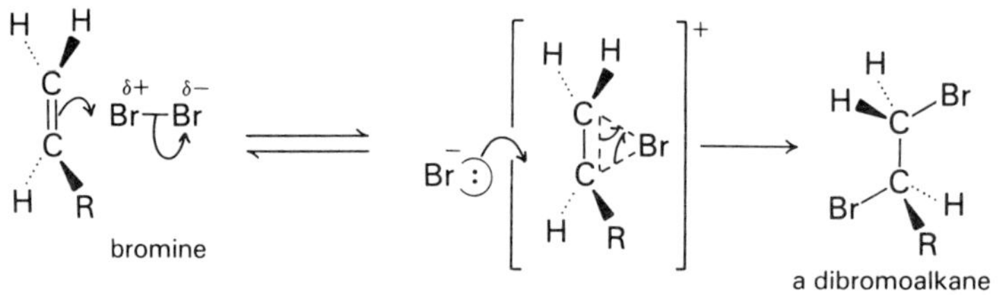

In each of the above reactions, a molecule is added across the carbon–carbon double bond.

Electrophilic substitution

Aromatic compounds undergo electrophilic substitution rather than addition. A substitution reaction retains the resonance stability associated with aromaticity, whereas addition leads to its loss. However, in the absence of activating effects (page 196), powerful electrophiles are needed to cause substitution. The two most inportant ones are nitryl cations and Friedel–Crafts electrophiles (chloroderivatives in the presence of a Lewis acid).

Nitryl ions. Benzene is converted to nitrobenzene by reaction with a mixture of concentrated nitric and sulphuric acids:

$$HNO_3 + 2H_2SO_4 \rightleftharpoons NO_2^+ + H_3O^+ + 2HSO_4^-$$

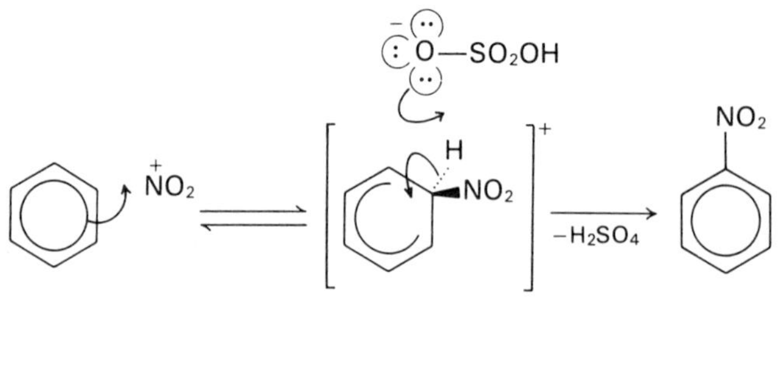

[Mechanism showing benzene reacting with NO_2^+ to form a Wheland intermediate, then loss of H_2SO_4 to give nitrobenzene.]

Friedel–Crafts reagents

In the reaction mechanism shown below, —R can be a chlorine atom, an alkyl group (e.g. —C_2H_5) or an alkanoyl group (e.g. —$COCH_3$). Electrophilic substitution occurs as shown below.

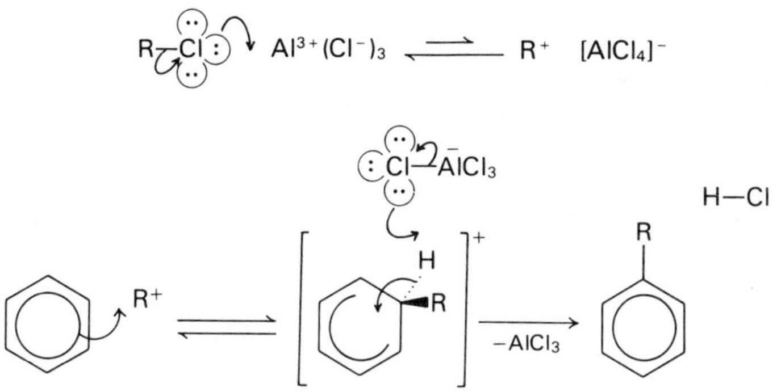

$$R-Cl: + Al^{3+}(Cl^-)_3 \rightleftharpoons R^+ \ [AlCl_4]^-$$

In each of the above reactions, substitution of a hydrogen atom on the ring happens. The activation energy for the reaction is likely to be high because the activated complex is not aromatic: ~160 kJ mol^{-1} are required to overcome the resonance energy of an aromatic system.

12 Classes of Organic Compounds

Functional groups and homologous series

The structure of a typical organic molecule contains a chain or a ring of bonded carbon atoms with other atoms (substituents) bonded on to the carbon skeleton. By considering the reactivity of a wide variety of these molecules, it is possible to establish different classes of compounds and, in each class, similarities in molecular structure become evident. These similarities take the form either of electron-deficient sites, or of electron-pairs able to attack electron-deficient sites. For example, the carbon atom bonded to a hydroxyl group in an alcohol molecule is electron-deficient. At the same time, however, the lone pairs of the hydroxyl group lead to nucleophilic properties as well.

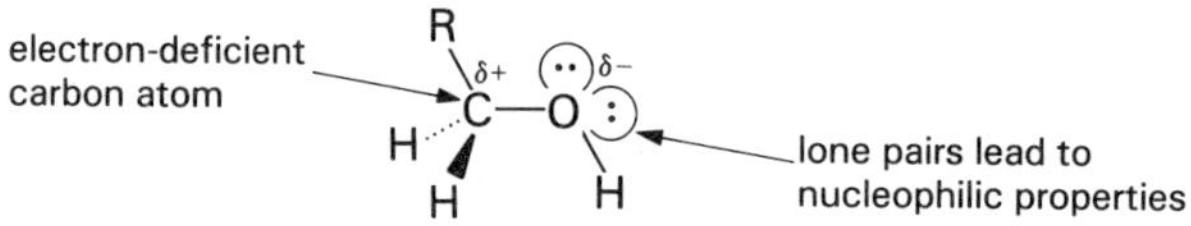

The differences between molecules of a particular class focus on the length of the carbon skeleton in each case. Because carbon–carbon and carbon–hydrogen bonds are strong, and so tend not to break under most reaction conditions, the carbon skeleton represents an unreactive part of the structure. The reactivity is, therefore, controlled by the particular substituents having electron-deficiency or electron-pair donor properties. A substituent which is the dominant factor in determining the reactivity of a compound is called its functional group.

> A functional group is a group of bonded atoms within a molecule which give the molecule its characteristic chemical properties.
> A homologous series is a class of compounds with the same functional group but whose molecules have a different number of —CH$_2$— links in their chains.

12.1 ALKANES

Structure and reactivity

Alkanes are saturated hydrocarbons of general formula C_nH_{2n+2}, and their functional group is best described as a carbon atom bonded either to hydrogen atoms or saturated carbon atoms. The molecules have a zig-zag structure with hydrogen atoms bonded at right angles to the direction of the carbon chain. Two examples are shown below.

a straight-chain hydrocarbon
(butane, C_4H_{10})

a branched-chain hydrocarbon
(methylpropane, C_4H_{10})

All the bond angles are close to 109° 31'. There are no electron-deficient sites and no reactive electron-pairs in the structure of an alkane. Also, because the bonds are strong (~400 kJ mol^{-1}), there is a lack of heterolytic reactivity. All the reactions of an alkane are homolytic in character.

Reactions

With chlorine. In the presence of sunlight, all organic compounds undergo homolytic chlorination. The methane-chlorine reaction is described on page 79.

With oxygen. The vast majority of organic compounds also react with oxygen via a homolytic process. The mechanism is more complex than that of chlorination because it involves a 'branched' chain reaction. The collision of a chain-propagating radical leads to the production of two radicals instead of just one. This has the effect of accelerating the chain reaction to an even more rapid rate. Alkanes burn readily in oxygen but explode with oxygen if a mixture of the two is sparked.

$$C_nH_{2n+2} + \left(\frac{3n+1}{2}\right)O_2 \longrightarrow nCO_2 + (n+1)H_2O$$

The reaction of any hydrocarbon with oxygen can be written in this sort of form. It is possible to work out the formula of an unknown hydrocarbon by measuring the volume of oxygen needed to burn a fixed volume of the gas. For example, consider the combustion of a hydrocarbon whose general formula is C_xH_y.

$$C_xH_y + (x + y/4)O_2 \longrightarrow xCO_2 + (y/2)H_2O$$

1. A measured volume of gaseous hydrocarbon (V_1) is mixed with an excess of oxygen of volume (V_2).
2. The mixture is ignited by spark and the volume is measured again when the gases have reached room conditions (V_3).
3. Finally, the gases produced are shaken with alkali, which absorbs the carbon dioxide and leaves only the excess oxygen in the vessel. This final volume (V_4) is recorded.
4. The following conclusions are drawn:

volume of hydrocarbon = V_1

volume of oxygen used = $(V_2 - V_4)$

volume of carbon dioxide = $(V_3 - V_4)$

Because the reaction is a gaseous one, the volume ratios are the same as the molar ratios (Avogadro's hypothesis), and, therefore, x and y are found as follows.

$$[C_xH_y]:[CO_2] = 1:x = V_1:(V_3 - V_4)$$

molar ratios   volume ratios

$$[C_xH_y]:[O_2] = 1:(x + y/4) = V_1:(V_2 - V_4)$$

$$\therefore x/1 = (V_3 - V_4)/V_1 \quad \text{and} \quad (x + y/4)/1 = (V_2 - V_4)/V_1$$

$$\therefore x = \frac{V_3 - V_4}{V_1} \quad \text{and} \quad y = \frac{4(V_2 - V_3)}{V_1}$$

For example, 10 cm³ of a gaseous hydrocarbon exploded in 125 cm³ of oxygen gives products of volume 110 cm³. After shaking with alkali, the volume drops to 50 cm³. By using the above working, it can be shown that the hydrocarbon is benzene, C_6H_6.

12.2 ALKENES

Structure, preparation and reactivity

The alkenes are unsaturated hydrocarbons containing carbon–carbon double bonds. A monoalkene has a general formula C_nH_{2n} but dienes and trienes also exist. The functional group of these compounds is C=C and they owe their reactivity to the donor properties (see page 185) of the π-electrons. For example, a monoalkene contains a single π-bonded link in its hydrocarbon chain:

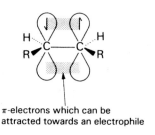

π-electrons which can be attracted towards an electrophile

Some dienes have 'conjugated' π-systems. Conjugation is the term given to the overlap effect illustrated below for butadiene.

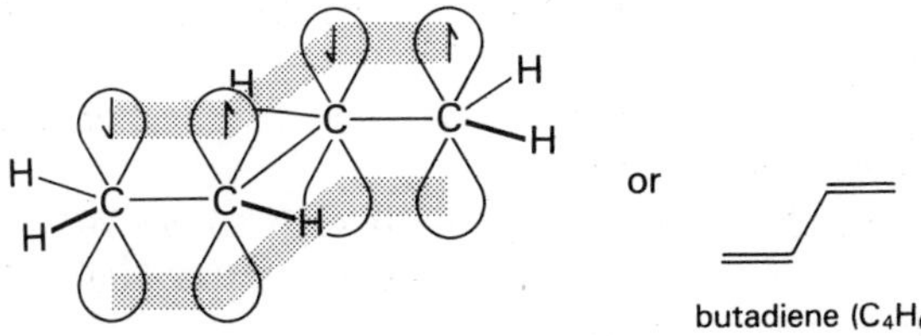

butadiene (C_4H_6)

When two π-bonded groups are joined by a σ-bond, the π-overlap spreads across this bond: the overall effect is called conjugation. Other examples of conjugated systems are seen on pages 201, 206 and 218; aromaticity (page 194) is a particularly important example of conjugation.

An alkene is best prepared by the dehydration of an alcohol, or from a halogenoalkane by elimination of hydrogen halide. For example, propene is obtained as follows:

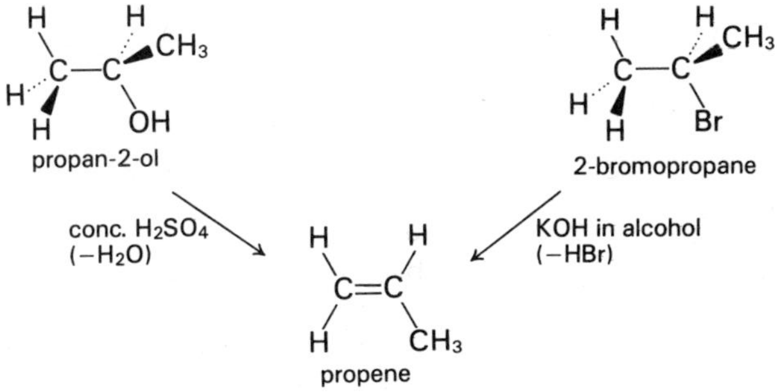

propan-2-ol   2-bromopropane

conc. H_2SO_4 (−H_2O)   KOH in alcohol (−HBr)

propene

Reactions

'*trans-*' electrophilic addition. Alkenes undergo electrophilic addition with aqueous halogens and with strong acids (see page 185). The mechanism is '*trans-*' in the sense that one addition occurs on one side of the molecule whereas the second takes place from the other side. Addition follows Markovnikov's rule which states that, on adding HX to an alkene molecule, the hydrogen atom adds to the carbon atom with the most hydrogen atoms already bonded there. The whole mechanism is discussed further on page 249; some examples of the reactions of propene are shown below and overleaf.

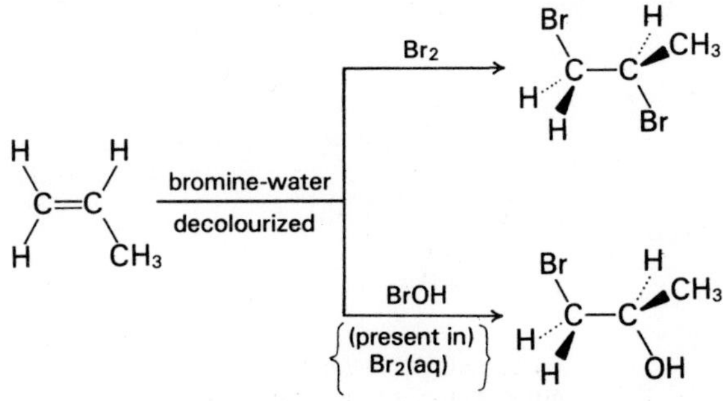

bromine-water decolourized

Br_2

BrOH (present in) $Br_2(aq)$

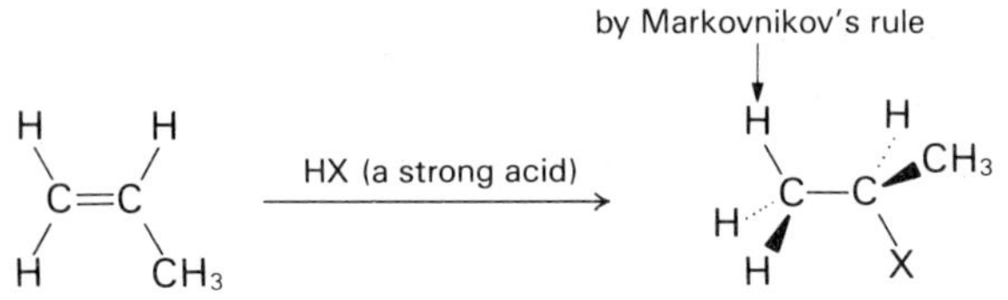
by Markovnikov's rule
HX (a strong acid)

'*cis-*' addition. Alkenes undergo '*cis-*' addition with hydrogen at a palladium or platinum surface at 150°C. The reaction takes place between 'chemisorbed' hydrogen atoms and adsorbed alkene molecules.

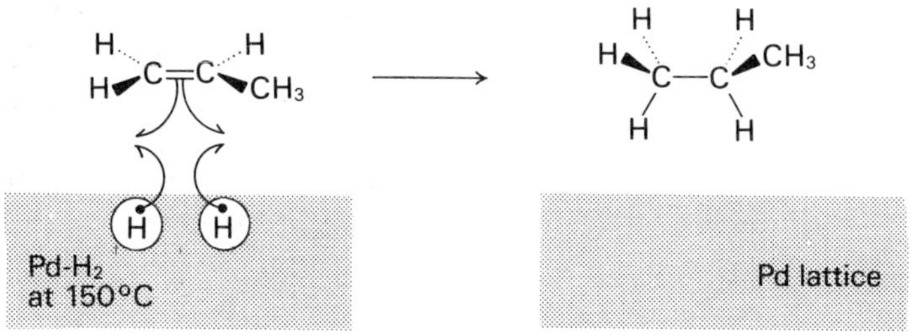
Pd-H2
at 150°C
Pd lattice

$$C_3H_6(g) + H_2(g) \xrightarrow[150°C]{Pd} C_3H_8(g)$$

Alkenes also react with alkaline manganate(VII) solution to give a *cis*- addition product; the purple colour of the manganate(VII) becomes a brown suspension of manganese dioxide.

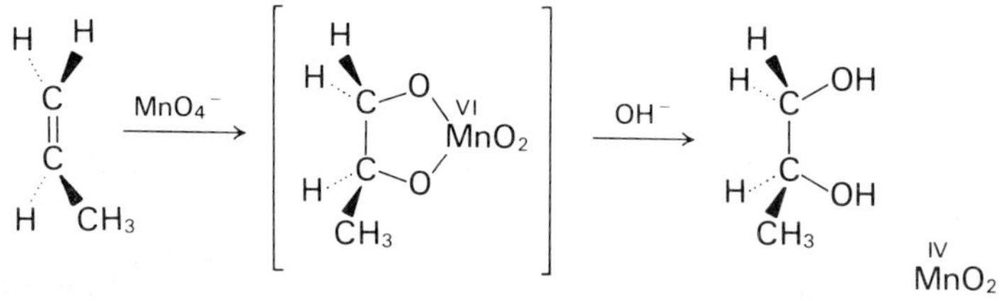
MnO4⁻
OH⁻
MnO2

Ozonolysis. Ozone (trioxygen) adds to an alkene; under mildly reducing conditions (zinc in ethanoic acid), the ozonide is broken down into aldehydes or ketones. One mole of ozone is used for every one mole of double bonds in the alkene.

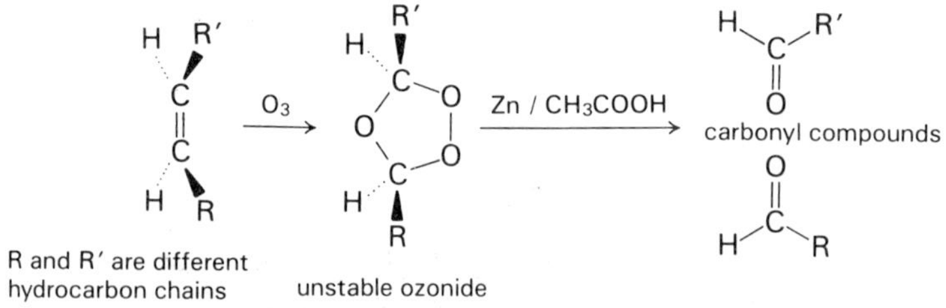
O3
Zn / CH3COOH
carbonyl compounds
R and R′ are different
hydrocarbon chains
unstable ozonide

The resultant carbonyl compounds can be separated by distillation and recognised from the melting-points of their oximes (see page 213). 'Ozonolysis' thus enables us to find out the number of double bonds per molecule in an unknown alkene, and in what position on the chain each double bond occurs. For example, a cycloalkene reacts with ozone in a molar ratio of 1:2 and there are only two products: ethanedial and butanedial. This shows that the alkene has two double bonds and must have the 1,3-hexagonal structure shown below.

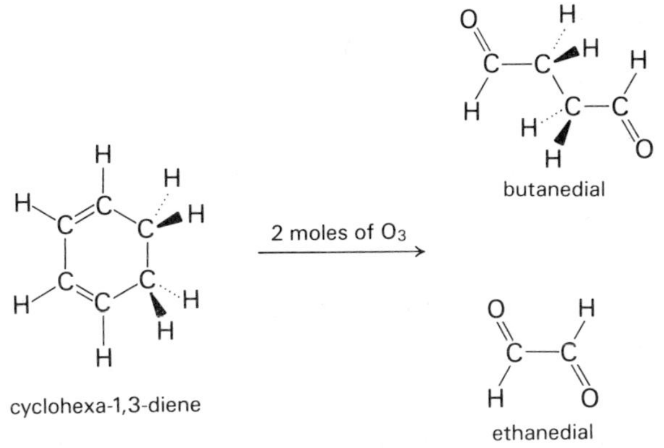

cyclohexa-1,3-diene → (2 moles of O_3) → butanedial + ethanedial

Polymerization. Alkenes polymerize in the presence of a free radical initiator at high pressure. For example, ethene gives poly(ethene) when mixed with a trace of ethanoyl peroxide at 150°C and 20 MPa.

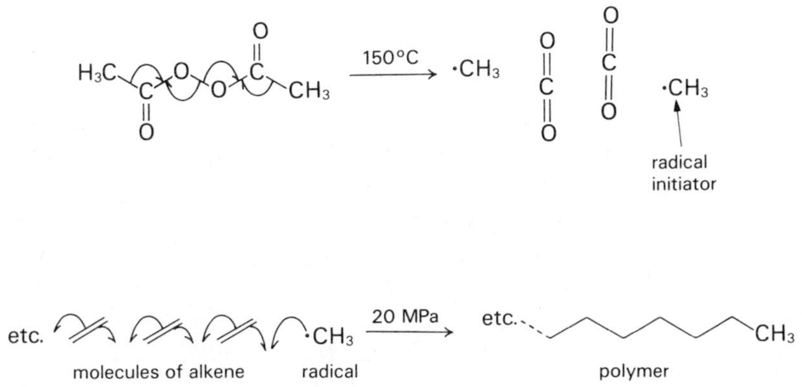

molecules of alkene + radical → (20 MPa) → polymer

Polymerization can be brought about at far lower pressures and temperatures with a Ziegler–Natta catalyst of titanium(III) chloride and triethylaluminium. The reaction takes place in a controlled heterolytic manner at each titanium ion, as shown overleaf.

$Al(C_2H_5)_3 + TiCl_3 \rightleftharpoons Al(C_2H_5)_2Cl + TiCl_2(C_2H_5)$

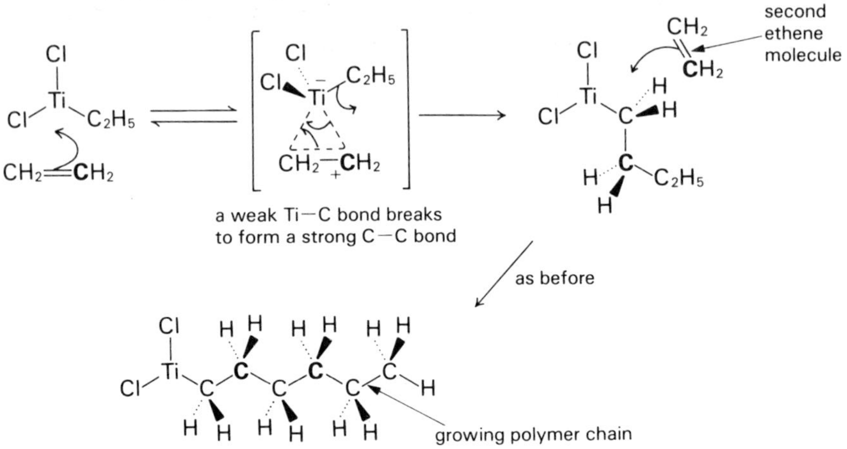

a weak Ti—C bond breaks to form a strong C—C bond

as before

second ethene molecule

growing polymer chain

12.3 AROMATICITY

Characteristics

Aromaticity is associated with a group of unsaturated ring compounds that have the following characteristics.

1 They often have lingering 'aromatic' smells.
2 They burn with a smoky, yellow flame producing a large proportion of soot.
3 Although they are unsaturated, they are unusually resistant to hydrogenation.
4 They do not undergo the addition reactions typical of alkenes (page 191).

The term aliphatic is used to describe any hydrocarbon (or its derivative) that does not have aromatic properties.

Structure

In all aromatic compounds, it is found that the molecules contain a cyclic σ-framework and a conjugated π-system. The necessary condition for aromaticity is that there should be a total of $(4n + 2)\pi$-electrons ($n = 0, 1, 2, 3\ldots$) delocalized over the whole cyclic structure. Many aromatic compounds are based on benzene, C_6H_6, whose structure contains 6 π-electrons ($n = 1$) delocalized over a hexagonal cyclic σ-framework.

The effect of the conjugation is best illustrated by the side-on view of a benzene molecule drawn below. For simplicity, the six hydrogen atoms have been omitted; they are bonded to each carbon atom in the plane of the hexagon.

the molecule contains six π-electrons delocalized above and below a hexagon of σ-bonded carbon atoms

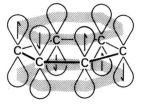

each carbon atom has a single-bonded hydrogen atom (not shown)

12 Classes of Organic Compounds

In a simple valence-bond drawing, benzene can only be represented as cyclohexatriene. There are two main canonical forms of this structure:

canonical forms of C_6H_6

the resonant hybrid

The increased bond energy that results from the effect of conjugation gives an aromatic ring a much greater stability than the valence-bond model suggests. The extra energy is called the resonance (or delocalization) energy, and can be shown by Born–Haber calculations (page 66) to be of the order of 160 kJ mol^{-1}.

It is interesting to note that cyclobutadiene and cyclo-octatetraene are *not* aromatic because they do not contain $(4n + 2)$ delocalized π-electrons. The first has four and the second eight, and in neither case is n a whole number. The three structures shown in the table below, however, all have aromatic properties.

$n=0$	$n=1$	$n=2$
$C_3H_3^+$, obtained from the molecule C_3H_3I by loss of an iodide ion	C_4H_4O; one of the lone pairs of electrons on the oxygen atom is delocalized over the ring	$C_{10}H_8$; two benzene rings fused together. There is no room for hydrogen atoms at the join, and so there is more carbon than hydrogen in the molecule

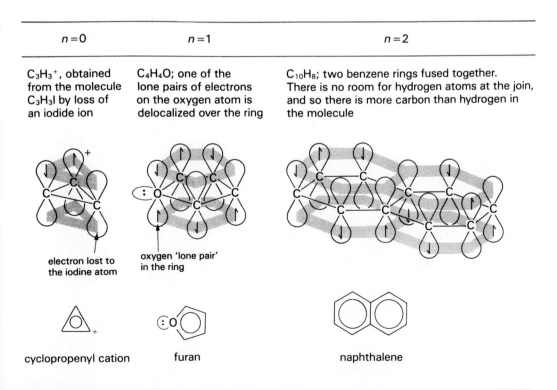

cyclopropenyl cation   furan   naphthalene

Reactivity

Aromatic compounds do not undergo addition or hydrogenation reactions readily because these lead to the disruption of aromaticity. It requires 160 kJ mol^{-1} to overcome the delocalization energy associated with the aromatic ring. Instead, electrophilic attack tends to produce a substitution product in which aromaticity is retained.

In many substituted aromatic molecules, further conjugation takes place. This has two major effects: it alters the ability of the π-electrons to be attracted to electrophiles, and it leads to the formation of stereoselective products (see page 247). The impact of a substituent on the reactivity of the ring can be summarized in terms of the mesomeric (M) and inductive (I) effects of the particular substitutent (see page 181). A substituent which is $+M$ or $+I$ donates electron density to the ring and, therefore, activates it for electrophilic attack. A substituent that is $-M$ or $-I$ deactivates the ring and makes the substituted compound less reactive than its parent compound.

	activating substituents		deactivating substituents
$+M$	—OH —NH$_2$	$-M$	—NO$_2$ —C=O
$+I$	—CH$_3$ —C$_2$H$_5$ etc.	$-I$	—Cl —Br etc.

Activating substituents tend to *direct* any further substitution to the 2-, 4- and 6- positions, whereas deactivating substituents tend to be 3-, 5- *directing*. This is discussed further on page 247.

For example, the reactivity of benzene, nitrobenzene and phenylamine towards the weakly electrophilic reagent, bromine, follows the order shown below.

activating   C$_6$H$_5$NH$_2$ > C$_6$H$_6$ > C$_6$H$_5$NO$_2$ deactivating

Also, the amino-group directs 2-, 4-, 6- whereas the nitro-group directs 3-, 5- and, therefore, the products are different.

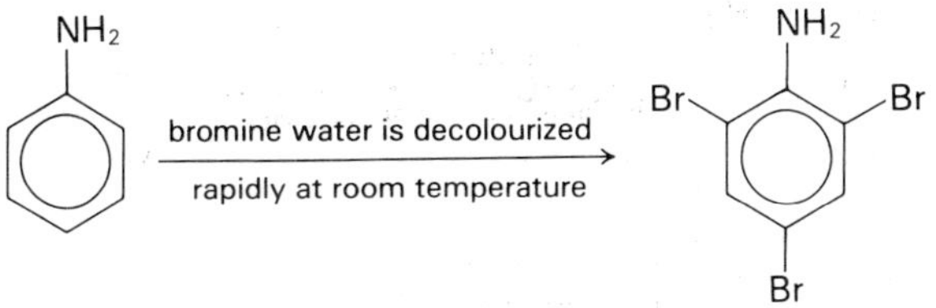

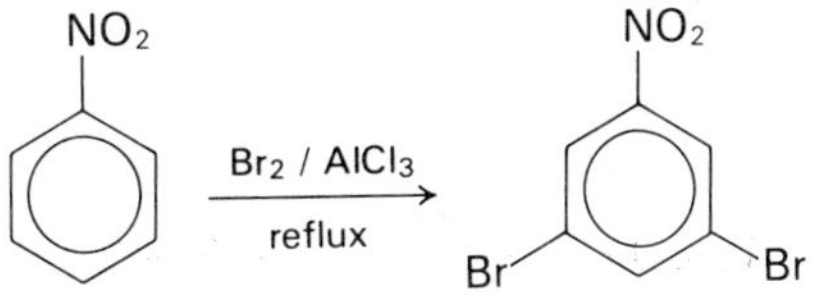

[Reaction scheme: aniline + bromine water → 2,4,6-tribromoaniline; bromine water is decolourized rapidly at room temperature]

[Reaction scheme: nitrobenzene + Br₂/AlCl₃, reflux → 3,5-dibromonitrobenzene]

12.4 HALIDES

Structure, preparation and reactivity

A halogen atom is one of the simplest of substituents. Monohalides and polyhalides containing up to three halogen atoms per carbon atom can be produced, for example:

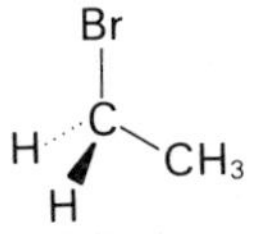

bromoethane 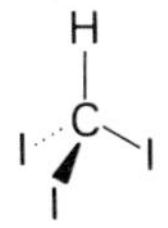   triiodomethane (iodoform) 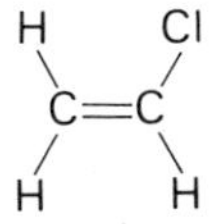   chloroethene (vinyl chloride) 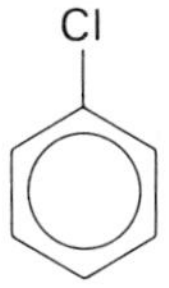   chlorobenzene

The presence of a halogen atom in a hydrocarbon molecule causes charge separation. The carbon atom bonded to the halogen becomes electron-deficient and susceptible to nucleophilic attack.

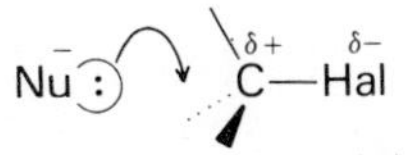

$$Nu:^- \quad \overset{\delta+}{C}-\overset{\delta-}{Hal}$$

Halides are prepared in a number of different ways. For chloro-derivatives the major ones are as follows.

From chlorine

$$CH_4(g) \xrightarrow[\text{ultraviolet light}]{Cl_2 \text{ in}} CH_3Cl, CH_2Cl_2, CHCl_3, CCl_4$$

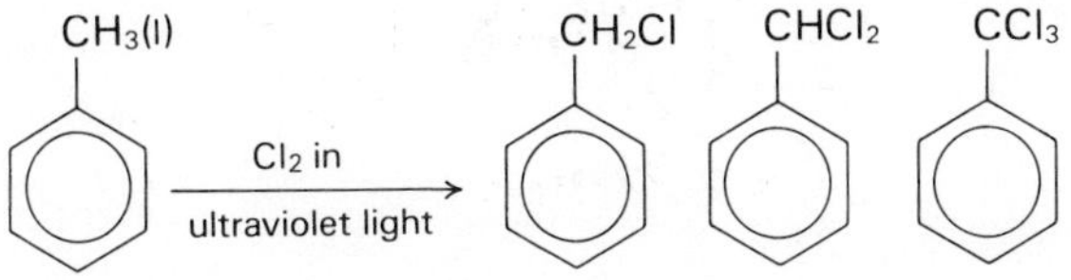

$$CH_3(l)\text{-C}_6H_5 \xrightarrow[\text{ultraviolet light}]{Cl_2 \text{ in}} \text{C}_6H_5\text{-}CH_2Cl, \text{C}_6H_5\text{-}CHCl_2, \text{C}_6H_5\text{-}CCl_3$$

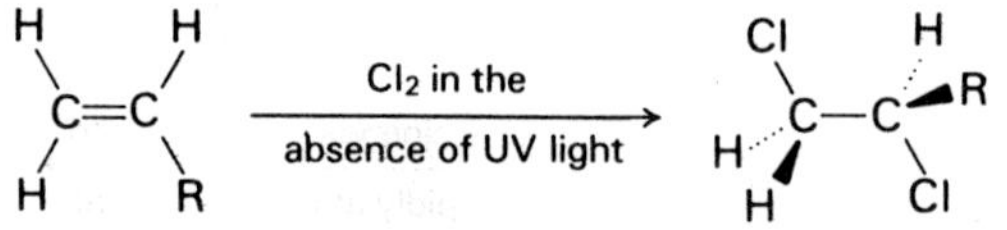

From chlorides

$$CH_2=CHR \xrightarrow{Cl_2 \text{ in the absence of UV light}} CH_2Cl-CHClR$$

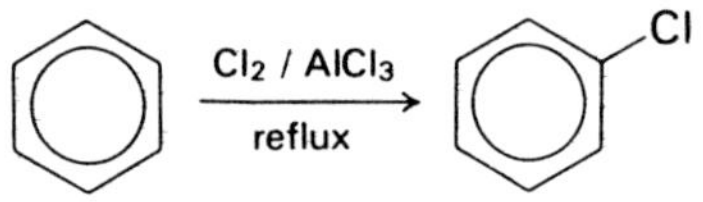

$$C_6H_6 \xrightarrow[\text{reflux}]{Cl_2 / AlCl_3} C_6H_5Cl$$

$$R-OH \xrightarrow{PCl_5} R-Cl$$

$$R-OH \xrightarrow[\text{2 KCl; reflux}]{1 \text{ conc. } H_2SO_4} R-Cl$$

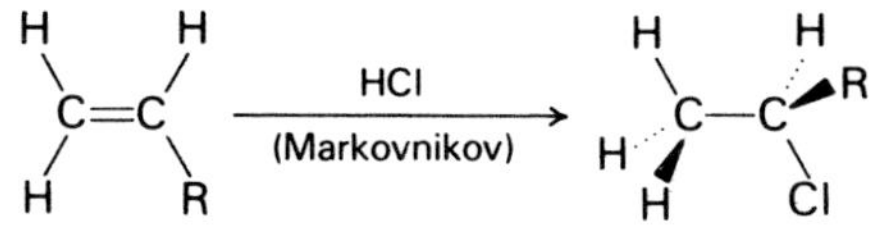

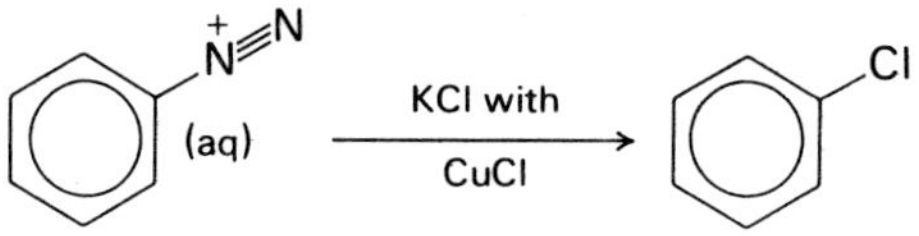

$$CH_2=CHR \xrightarrow[\text{(Markovnikov)}]{HCl} CH_3-CHClR$$

$$C_6H_5N_2^+ \text{ (aq)} \xrightarrow[\text{CuCl}]{KCl \text{ with }} C_6H_5Cl$$

Reactions

With nucleophiles. A saturated carbon atom bonded to a halogen atom can successfully be attacked by ammonia molecules, cyanide ions or hydroxide ions. The mechanisms of these reactions (S_N2 or S_N1) are discussed on page 182. For example,

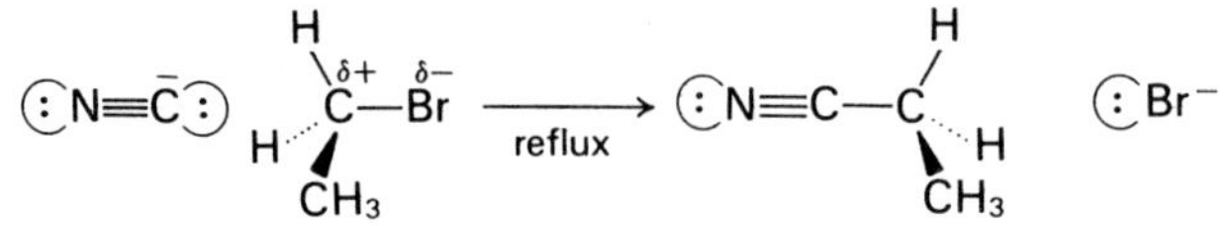

$$:N\equiv\bar{C}: \; + \; CH_3CHBrH \xrightarrow{\text{reflux}} :N\equiv C-CH(CH_3)H \; + \; :Br^-$$

In the case of the ammonia reaction, the initial product is a primary amine which is also a nucleophile. Further nucleophilic attack of the halogenoalkane can then occur so that primary, secondary and tertiary amines are produced, and finally a quarternary ammonium salt.

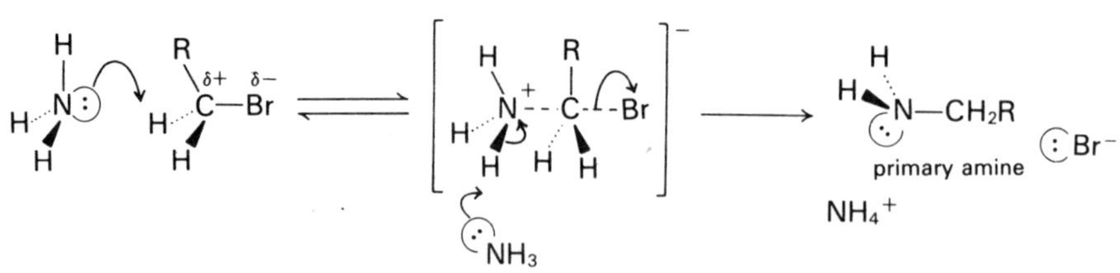

$$H_3N: \; + \; RCHBrH \; \rightleftharpoons \; [H_3N^+-CHRH \cdots Br]^- \xrightarrow{NH_3} H_2N-CH_2R \; + \; :Br^- \; + \; NH_4^+$$

primary amine

Then,

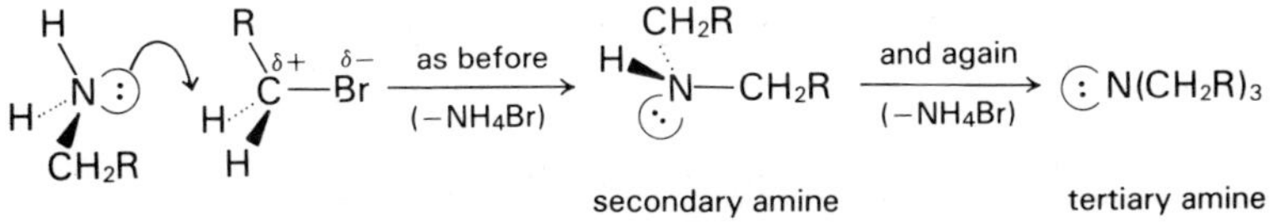

[Reaction scheme: H-N(H)(CH₂R) with H-C(R)(H)-Br (δ+, δ−) → (as before, −NH₄Br) → H-N(CH₂R)-CH₂R (secondary amine) → (and again, −NH₄Br) → :N(CH₂R)₃ (tertiary amine)]

An unsaturated carbon atom bonded to a halogen atom is less susceptible to nucleophilic attack. The electron-deficiency of the carbon atom is delocalized by a shift of π-electrons towards it. For example, aromatic halides are inert to alkali, even boiling alkali. Chlorobenzene is, in fact, a remarkably unreactive compound: the chlorine atom deactivates the ring while the aromatic π-electrons make nucleophilic attack on the halogen-bearing carbon atom unlikely.

With bases. Halogenoalkanes and halogenoalkenes undergo base-induced elimination of a hydrogen halide providing the base is strong enough, and does not act as a nucleophile. Typical bases include potassium hydroxide dissolved in ethanol, and sodium ethoxide. In the first case, the nucleophilic tendencies of the hydroxide ions are minimized by the effect of the bulky solvent molecules. These attendant solvent molecules make it more difficult for the lone pairs of electrons to reach the electron-deficient carbon atoms. Instead, the reaction shown below takes place.

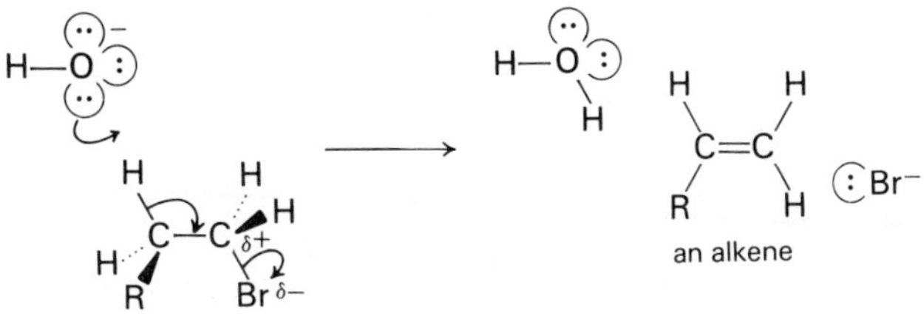

[Reaction mechanism: H-O⁻ attacks H on C-C bearing Br (δ+), eliminating to give H-O-H, an alkene (C=C with R, H, H, H substituents) and :Br⁻]

In the second case, when a solution of ethoxide is refluxed with 1-bromopropane, a mixture of propene (the elimination product) and 1-ethoxypropane (the substitution product) is obtained.

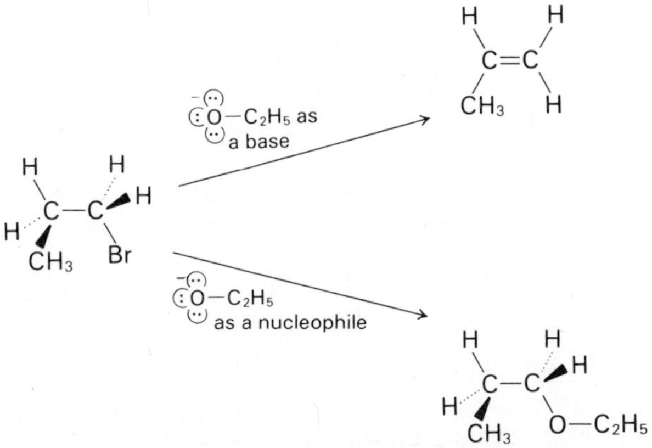

[Reaction scheme showing CH₃-CH(Br)-CH₂H (with H's) reacting with:
- ⁻:O-C₂H₅ as a base → CH₃-CH=CH-H (propene)
- ⁻:O-C₂H₅ as a nucleophile → CH₃-CH(H)-CH₂-O-C₂H₅]

12.5 ALCOHOLS AND PHENOLS

Structure, preparation and reactivity

An alcohol is a substituted hydrocarbon containing a hydroxyl group as a substituent. However, if the substituted hydrogen atom is from an aromatic ring, the compound is known as a phenol. Some examples are shown below.

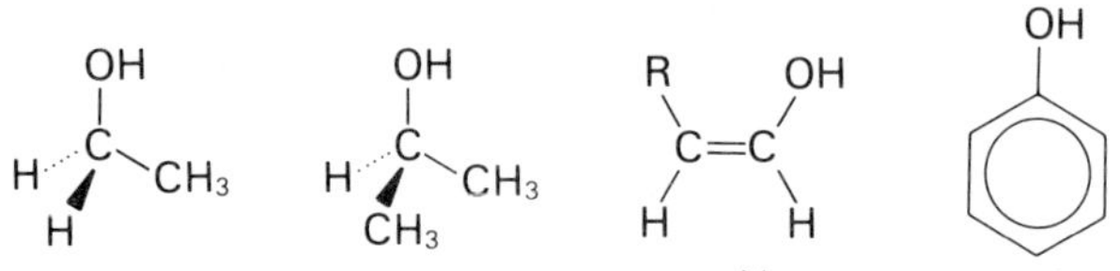

primary alcohol, secondary alcohol, an *enol* (see page 211), phenol

The terms primary, secondary and tertiary refer to the number of carbon atoms that bond to the alcohol functional group, C—OH. These terms are also used to classify halogenoalkanes and amines whose functional groups are shown below.

Alcohols are prepared by the hydrolysis of alkenes and halogenoalkanes, or by the reduction of any acid derivative. A primary alcohol can also be prepared by treating a primary amine with nitrous [nitric(III)] acid.

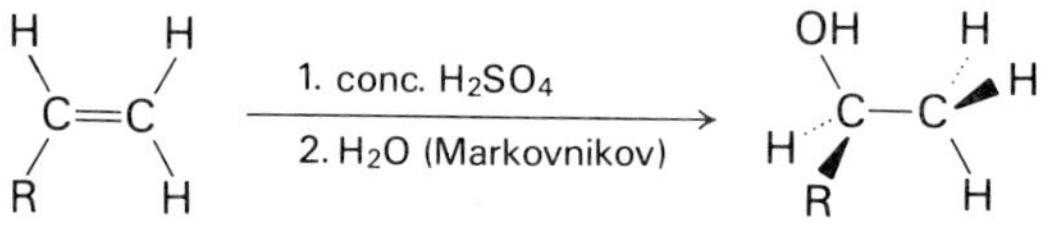

$$\text{C=C} \xrightarrow[\text{2. H}_2\text{O (Markovnikov)}]{\text{1. conc. H}_2\text{SO}_4} \text{C-C}$$

$$\text{R-Cl} \xrightarrow[\text{reflux}]{\text{OH}^-\text{(aq)}} \text{R-OH}$$

$$\underset{X}{\overset{R}{\text{C=O}}} \xrightarrow[\text{2. H}_2\text{O}]{\text{1. LiAlH}_4 \text{ in dry ether}} \text{R-CH}_2\text{OH}$$

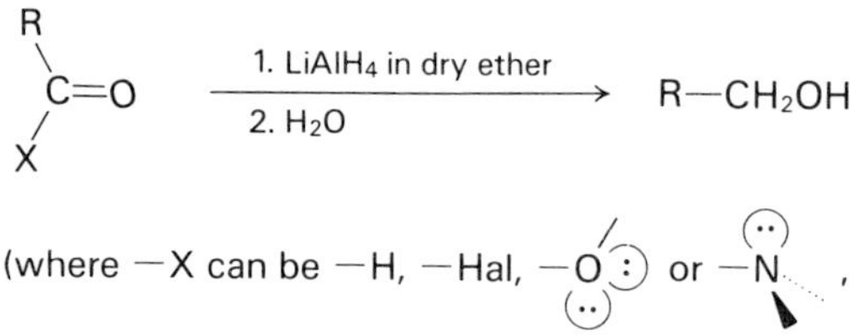

(where —X can be —H, —Hal, —O or —N ,

$$\text{R-NH}_2 \xrightarrow{\text{HONO}} \text{R-OH}$$

The reactivity of an alcohol is more complex than that of a halogenoalkane. Although the bonded oxygen atom exerts a $-I$ effect in the same way as a halogen atom, the hydroxyl group is less likely to be a leaving group than a halide. For nucleophilic attack to be successful, the hydroxyl group must usually be protonated first:

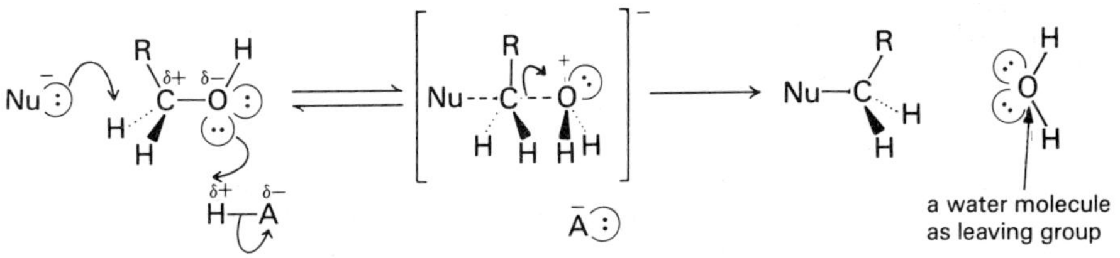

12 Classes of Organic Compounds

[Reaction scheme showing nucleophilic substitution with water as leaving group]

Also, the prominence of the lone pairs of the hydroxyl group gives the alcohol itself nucleophilic and basic properties.

alcohol as nucleophile   alcohol as base

In this respect an alcohol molecule resembles a water molecule.

Phenols cannot be prepared in the same way as alcohols. Two laboratory methods are shown in outline below for an unsubstituted phenol.

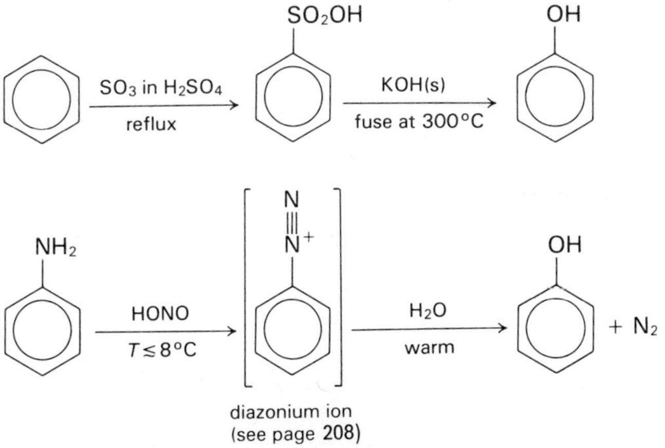

[Scheme 1: benzene → (SO$_3$ in H$_2$SO$_4$, reflux) → benzenesulfonic acid (SO$_2$OH) → (KOH(s), fuse at 300°C) → phenol (OH)]

[Scheme 2: aniline (NH$_2$) → (HONO, $T \leq 8$°C) → diazonium ion (see page 208) → (H$_2$O, warm) → phenol (OH) + N$_2$]

Phenol is manufactured industrially using the cumene process (page 175).

The reactivity of a phenol differs from that of an alcohol. Firstly, phenols are inert to nucleophilic attack in the same way that aryl halides are (page 199) and secondly, the properties of the hydroxyl lone pair are considerably altered by conjugation.

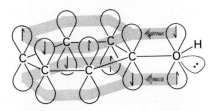

The lone pair strongly activates the aromatic ring by exerting a +M effect to offset the −I effect of the oxygen atom. For example, phenol reacts with cold dilute bromine water:

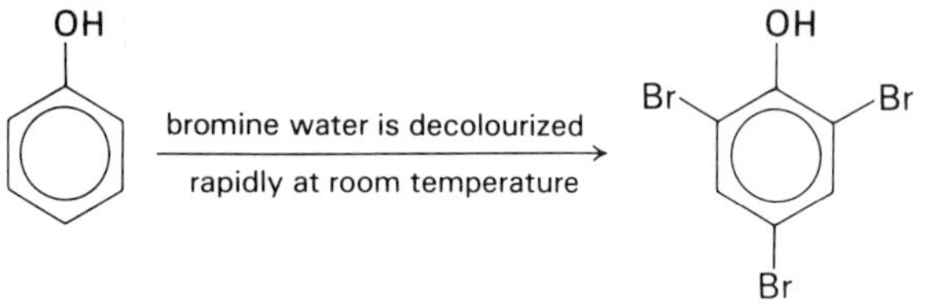

$$\text{C}_6\text{H}_5\text{OH} \xrightarrow{\text{bromine water is decolourized rapidly at room temperature}} \text{2,4,6-tribromophenol}$$

Also, the conjugation of the lone pair causes a phenol to be a weaker nucleophile and base than an alcohol. Conversely, phenols are stronger acids than alcohols because the charge of the anion is more readily delocalized by the conjugation effect.

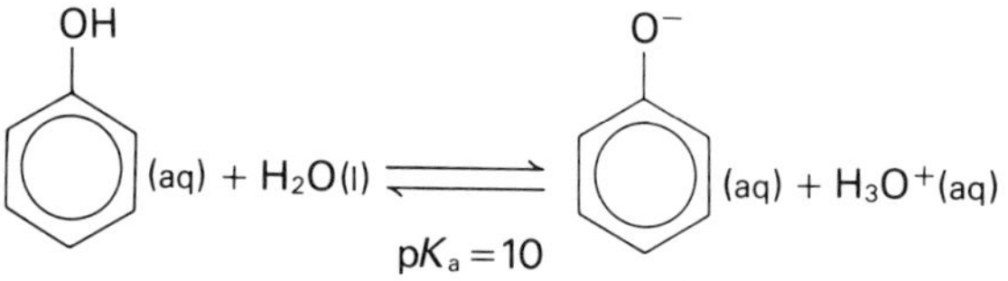

$$\text{C}_6\text{H}_5\text{OH(aq)} + \text{H}_2\text{O(l)} \rightleftharpoons \text{C}_6\text{H}_5\text{O}^-\text{(aq)} + \text{H}_3\text{O}^+\text{(aq)} \quad pK_a = 10$$

$$\text{C}_2\text{H}_5\text{OH(aq)} + \text{H}_2\text{O(l)} \rightleftharpoons \text{C}_2\text{H}_5\text{O}^-\text{(aq)} + \text{H}_3\text{O}^+\text{(aq)} \quad pK_a = 16$$

Reactions

With concentrated sulphuric acid. Alcohols are dehydrated to alkenes when treated with concentrated sulphuric acid. Primary alcohols are converted first to alkyl hydrogensulphates which decompose on heating to give alkenes. Secondary alcohols, however, only need to be warmed with the acid, while tertiary alcohols are even more readily dehydrated still. (Alternatively, alcohols can be dehydrated in the vapour phase over hot aluminium oxide.)

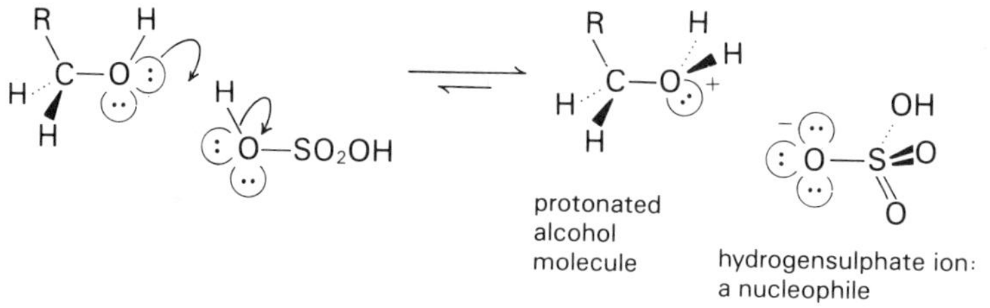

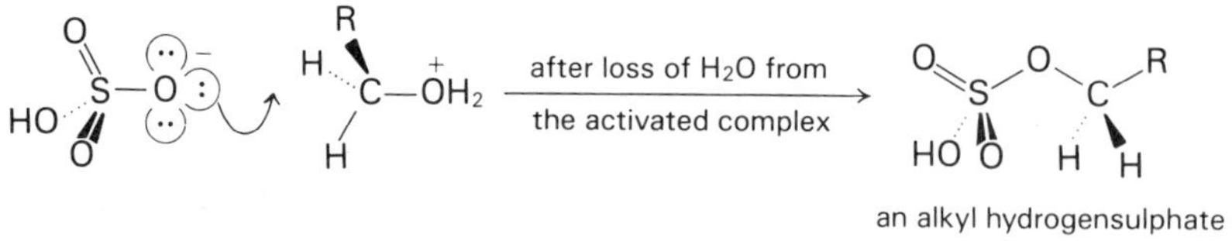

Alkyl hydrogensulphates decompose as shown on next page for the ethyl derivative.

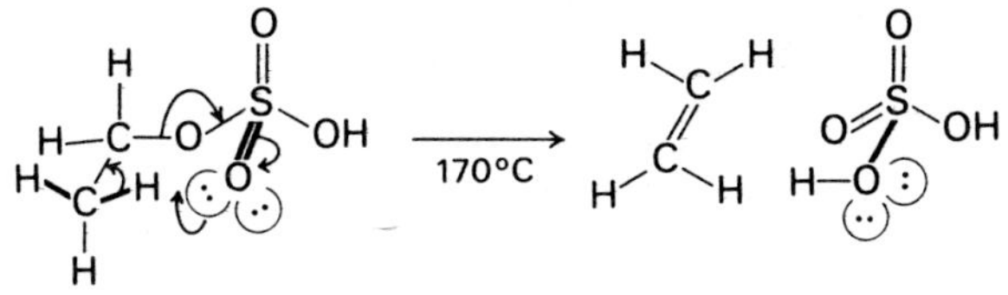

[Diagram: ethyl hydrogensulphate → ethene + sulphuric acid at 170°C]

If there is excess alcohol present in the reaction mixture, an ether is also produced as a result of the action of the excess alcohol as a nucleophile.

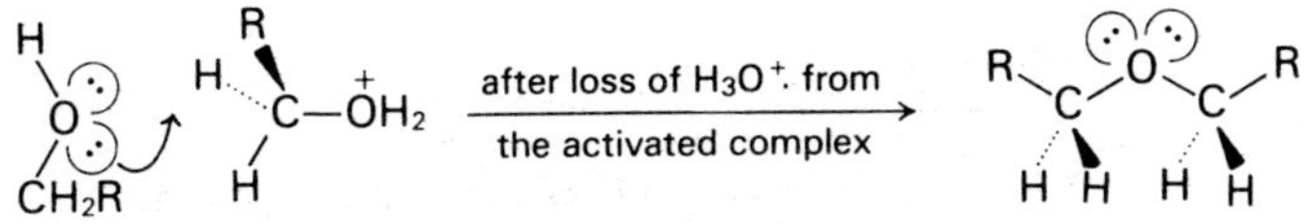

[Diagram showing nucleophilic attack of alcohol on protonated alcohol, after loss of H_3O^+ from the activated complex, giving $R-CH_2-O-CH_2-R$ type ether]

For ethanol, the sequence of reactions with concentrated sulphuric acid can be summarized as shown below.

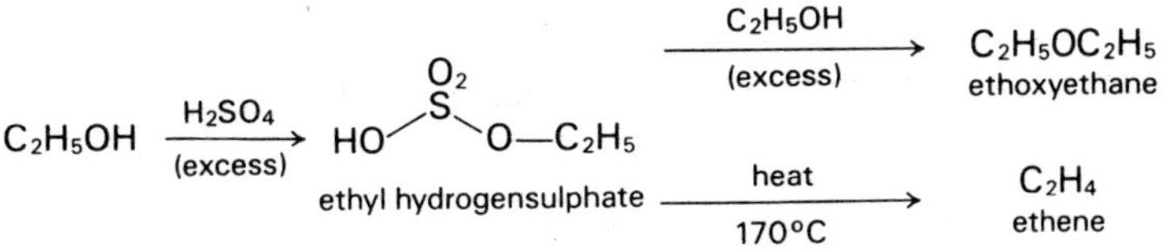

$$C_2H_5OH \xrightarrow[\text{(excess)}]{H_2SO_4} \text{HO-SO}_2\text{-O-C}_2H_5 \text{ (ethyl hydrogensulphate)}$$

$$\xrightarrow{C_2H_5OH \text{ (excess)}} C_2H_5OC_2H_5 \text{ ethoxyethane}$$

$$\xrightarrow[170°C]{\text{heat}} C_2H_4 \text{ ethene}$$

For a typical secondary alcohol, the dehydration product is the only one to be isolated from the reaction mixture.

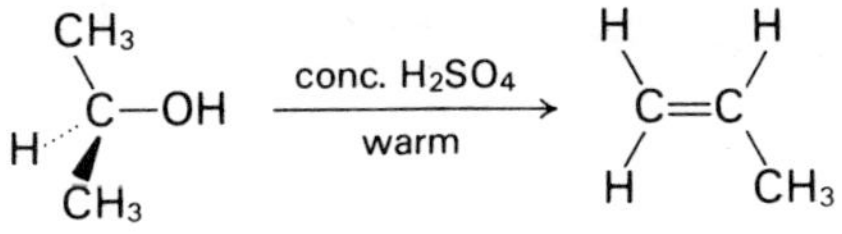

[Diagram: (CH_3)_2CH-OH → CH_3-CH=CH_2 with conc. H_2SO_4, warm]

With nucleophiles. The hydroxyl group does not often act as a leaving group, and so alcohols do not interact extensively with nucleophiles. By protonating this group using concentrated sulphuric acid, the reactivity towards nucleophiles can be increased. However, there are so many side-reactions that the process has little value. For example,

$$R-OH \xrightarrow[\text{2 KBr reflux}]{\text{1 conc. H}_2SO_4} R-Br \quad \text{yield } \sim 30\%$$

It is not possible to carry out a similar reaction using nitrile ions: these have sufficient basic strength to be protonated by the acid added. The following sequence must be employed instead:

$$R-OH \xrightarrow{PCl_5} R-Cl \xrightarrow[\text{reflux}]{KCN} R-CN$$

As a nucleophile. Alcohols and, to a lesser extent, phenols act as nucleophiles. In the presence of concentrated sulphuric acid, an alcohol reacts with a carboxylic acid (or another alcohol, as on page 203). The product is an ester, but the reaction does not go to completion.

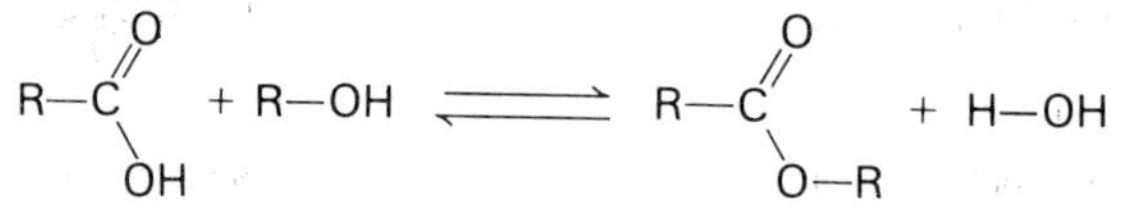

$$R-\underset{OH}{\underset{|}{C}}{=}O + R-OH \rightleftharpoons R-\underset{O-R}{\underset{|}{C}}{=}O + H-OH$$

Both alcohols and phenols attack an acid derivative more readily; these reactions are much quicker and go to completion. For example, the phenol ester, phenyl ethanoate can be prepared as follows.

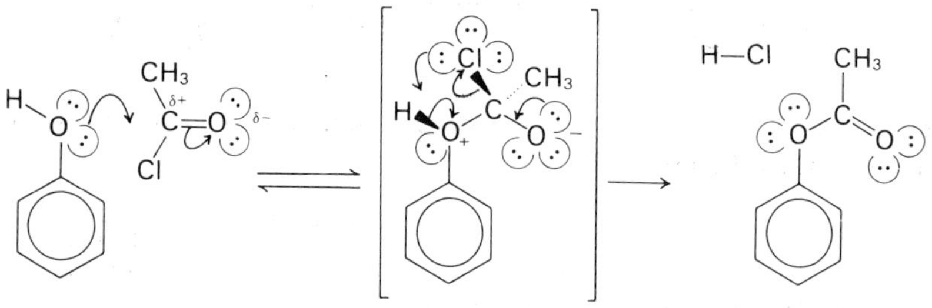

Alcohols attack phosphorus pentachloride or sulphurdichloride oxide to produce chloroalkanes; phenols do not react in this way.

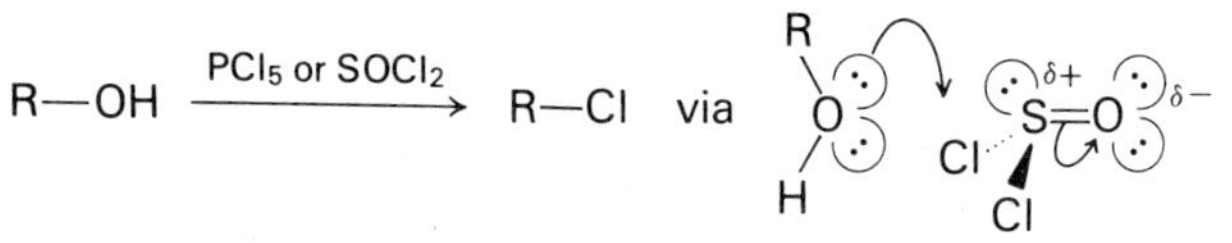

$$R-OH \xrightarrow{PCl_5 \text{ or } SOCl_2} R-Cl \text{ via } \dots$$

As a ligand. Alcohols and phenols form a range of complexes, mostly with transition metal cations. In particular, phenol forms a purple complex with iron(III) and this complex is a useful diagnostic test for phenolic compounds.

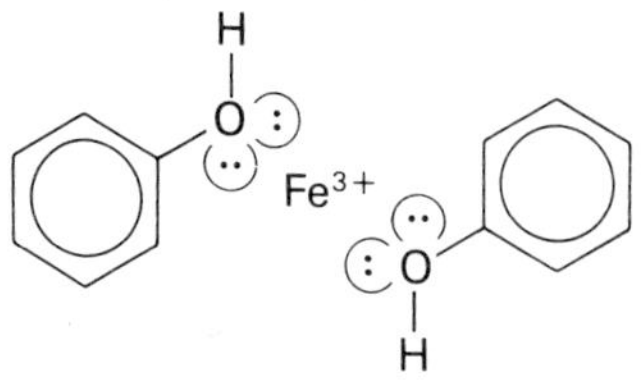

purple

As an acid. Phenols are stronger acids than alcohols (see page 202), but both are strong enough to react with sodium to give salts and hydrogen.

$$2R-OH + Na \longrightarrow 2R-O^- Na^+ + H_2$$

Phenol dissolves in sodium carbonate solution:

$$\text{C}_6\text{H}_5\text{OH (s)} + \text{CO}_3^{2-} \text{(aq)} \rightleftharpoons \text{C}_6\text{H}_5\text{O}^- \text{(aq)} + \text{HCO}_3^- \text{(aq)}$$

However, it is not strong enough to liberate carbon dioxide from this reaction mixture.

With oxidants. A primary alcohol is oxidized to a carboxylic acid by dichromate(VI) or manganate(VII). It is possible to isolate an aldehyde as an intermediate in this reaction.

$$\text{R}-\text{CH}_2\text{OH} \xrightarrow{\text{Cr}_2\text{O}_7^{2-}} [\text{R}-\text{CHO}] \xrightarrow{\text{Cr}_2\text{O}_7^{2-}} \text{R}-\text{COOH}$$

A secondary alcohol is oxidized to a ketone, but tertiary alcohols and phenols are oxidized only with great difficulty.

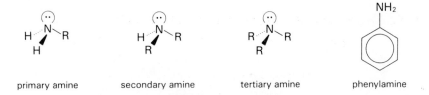

12.6 AMINES

Structure, preparation and reactivity

The structure of an amine is based on a saturated trivalent nitrogen atom. A primary amine has one carbon chain bonded to the nitrogen atom, a secondary amine has two chains and a tertiary amine has three. Aromatic amines bear the same relationship to aliphatic amines as phenols do to alcohols.

primary amine   secondary amine   tertiary amine   phenylamine

The reactivity of an aliphatic amine is dominated by the effects of the nitrogen lone pair. Although the nitrogen atom exerts a $-I$ effect on the carbon atom to which it bonds, the amino group is an even worse leaving group than a hydroxyl group. The reactivity of halogenoalkanes, alcohols and amines, therefore, follow the trends on page 206.

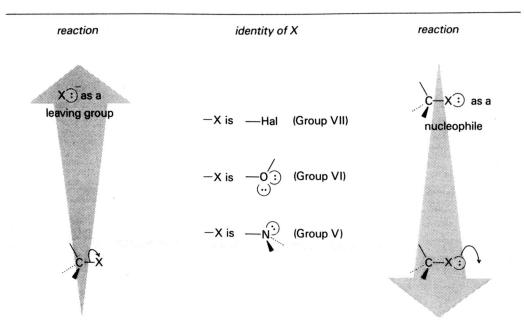

Halogenoalkanes do not readily act as nucleophiles, whereas alcohols show some reactivity both as nucleophiles and in their susceptibility to nucleophilic attack as well. Under aqueous conditions, however, an amine acts only as a nucleophile. The amino group activates an aromatic ring as powerfully as a hydroxyl substituent. The $+M$ effect easily overcomes the $-I$ effect exerted by the nitrogen atom.

For example, phenylamine decolorizes cold bromine water.

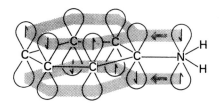

As in the case of phenols, the conjugation of the lone pair causes a considerable weakening of nucleophilic and basic strength.

Amines are prepared by hydrogenating nitriles, or by the Hofmann degradation of an amide. The reaction of ammonia (or other amines) with a halogenoalkane produces a mixture of amines and derived amines (see page 198). Aryl amines are prepared by the reduction of a nitro compound.

$$R-C\equiv N \xrightarrow[\text{(or } H_2-Pd \text{ at } 150°C)]{1. \text{ LiAlH}_4 \quad 2. \text{ H}_2\text{O}} R-CH_2NH_2$$

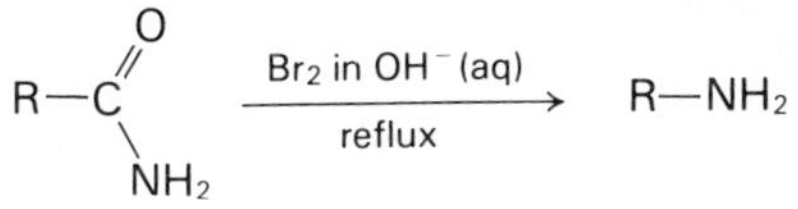

$$R-\overset{\displaystyle O}{\underset{\displaystyle NH_2}{\overset{\displaystyle \|}{C}}} \xrightarrow[\text{reflux}]{Br_2 \text{ in OH}^- \text{(aq)}} R-NH_2$$

$$R-Cl \xrightarrow[\text{(or RNH}_2 \text{ etc.)}]{NH_3} R-NH_2, R-NHR, (R)_3N \text{ etc.}$$

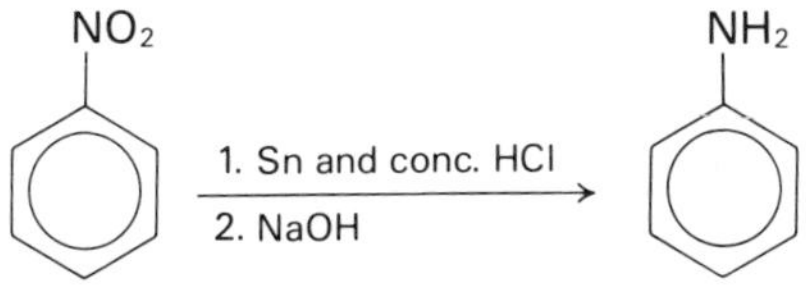

$$C_6H_5NO_2 \xrightarrow[\text{2. NaOH}]{\text{1. Sn and conc. HCl}} C_6H_5NH_2$$

Reactions

As a nucleophile. Amines react with halogenoalkanes and with acid derivatives: substituted amines and amides are produced respectively.

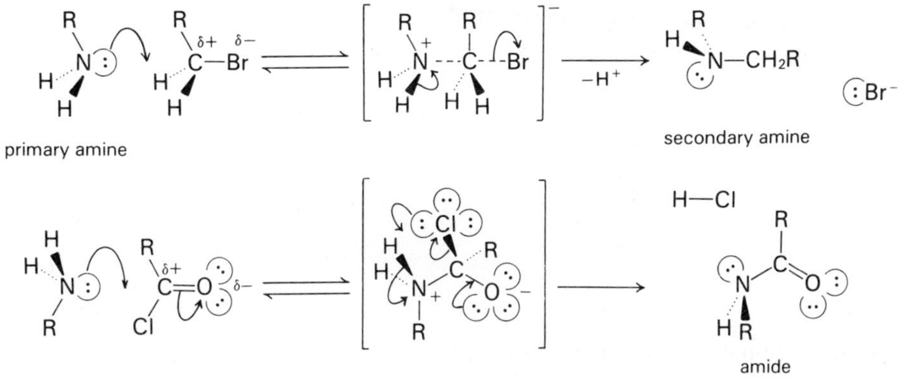

[Mechanism diagram: primary amine reacting with R—C—Br to form secondary amine + :Br⁻, with loss of H⁺]

[Mechanism diagram: amine reacting with acyl chloride (C=O with Cl) via tetrahedral intermediate to form amide + HCl]

As a base. An amine reacts with an acid to give a substituted ammonium salt; for example, dimethylamine neutralizes ethanoic acid to give dimethylammonium ethanoate.

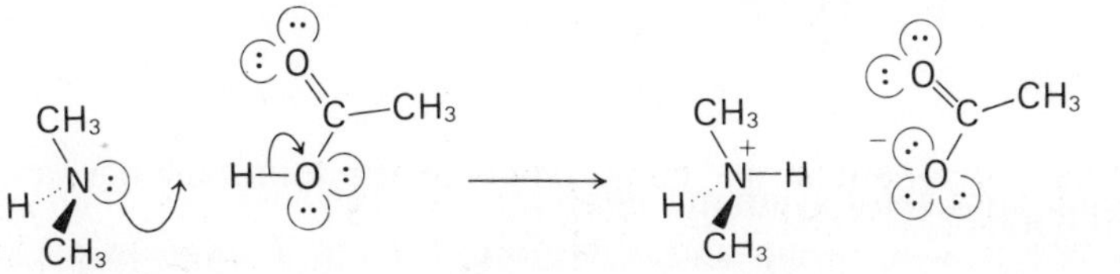

[Reaction diagram: (CH₃)₂NH + CH₃COOH → (CH₃)₂NH₂⁺ CH₃COO⁻]

A quaternary ammonium salt is produced when a tertiary amine is treated with a halogenoalkane. These salts undergo Hofmann elimination of an alkene when refluxed in alkali. For example, ethene and triethylamine are produced when tetraethylammonium chloride, $[N(C_2H_5)_4]^+ Cl^-$ is refluxed in alkali.

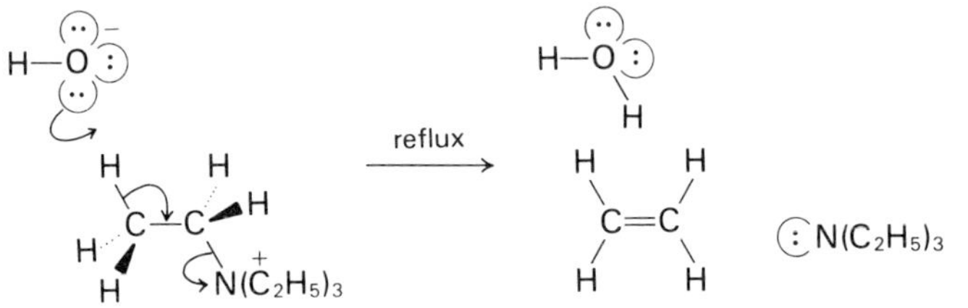

As a ligand. Amines form complexes in the same way that ammonia does. Two well-known examples are shown below.

$Cu(RNH_2)_4^{2+}$ $Cu(C_6H_5NH_2)_2^{2+}$

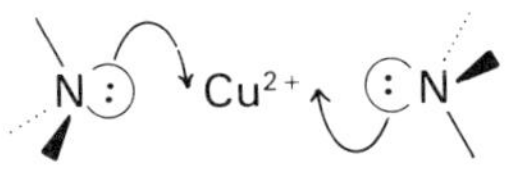

dark blue   yellow–green

With nitrous [nitric(III)] acid. The different classes of amine undergo different reactions with nitrous acid. Primary amines give alcohols and nitrogen, secondary amines produce insoluble, oily nitrosamines and tertiary amines form soluble nitrite salts.

$$R-NH_2 + HONO \longrightarrow R-OH + N_2 + H_2O$$
$$R-NHR + HONO \longrightarrow O=N-N(R)_2 + H_2O$$
$$R-N(R)_2 + HONO \longrightarrow H-N(R)_3^+ NO_2^-$$

The reactions can be used to distinguish between each class. Aromatic amines react with nitrous acid to give a solution containing diazonium ions, providing the temperature of the system is kept below 8°C.

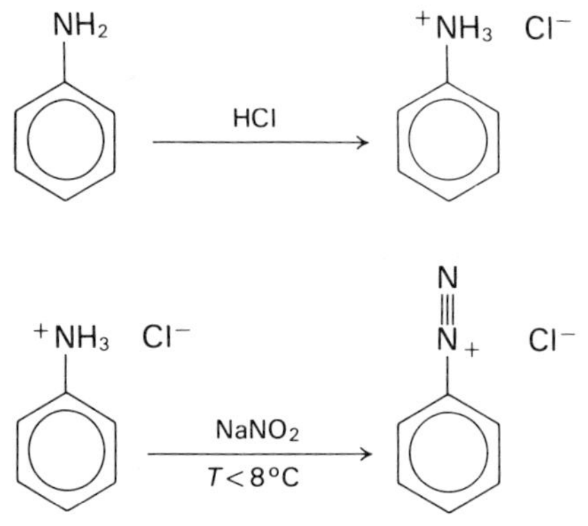

benzenediazonium ion

Diazonium ions are useful synthetic intermediates because the dinitrogen substituent tends to act as a leaving group and, therefore, the ion is open to nucleophilic attack.

12 Classes of Organic Compounds

Since aromatic reactions are mostly electrophilic in character, the diazonium ion provides a significant alternative set of reactions. Some examples are given below.

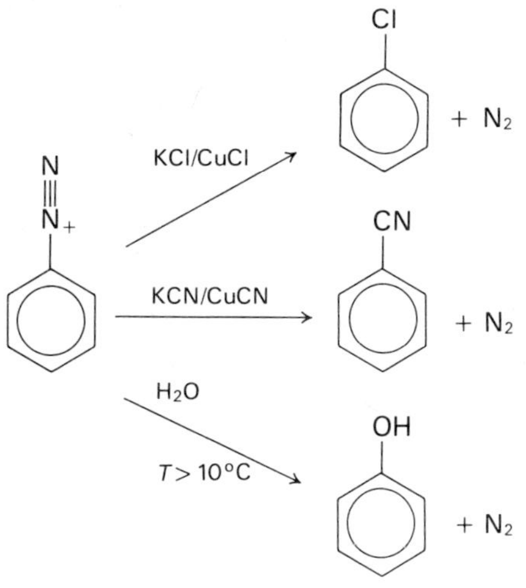

PhN₂⁺ → (KCl/CuCl) → chlorobenzene + N_2

PhN₂⁺ → (KCN/CuCN) → benzonitrile + N_2

PhN₂⁺ → (H_2O, $T > 10°C$) → phenol + N_2

Diazonium ions can also be reduced in one of two ways. The first retains the nitrogen atoms and the second leads to their loss.

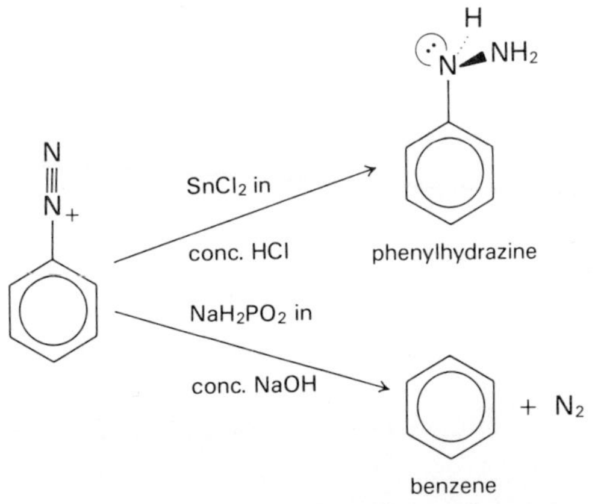

PhN₂⁺ → ($SnCl_2$ in conc. HCl) → phenylhydrazine (Ph–NH–NH₂)

PhN₂⁺ → (NaH_2PO_2 in conc. NaOH) → benzene + N_2

The second method provides a means by which 1,3,5-tribromobenzene can be produced. This compound is difficult to prepare from bromobenzene because the bromine substituent is not only deactivating, but also directs 2,4,6. Instead of starting with bromobenzene, phenylamine is used as follows.

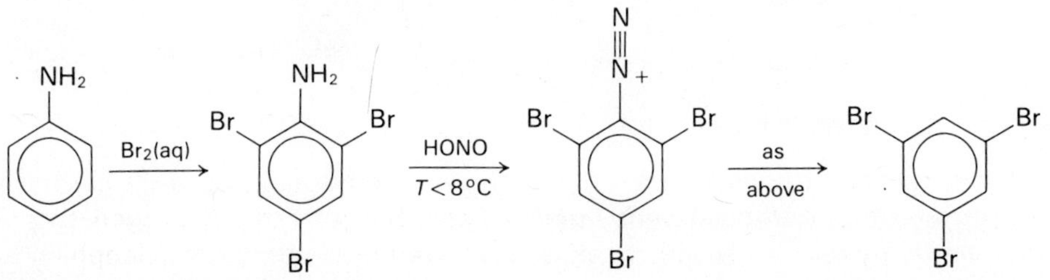

PhNH₂ → (Br_2(aq)) → 2,4,6-tribromoaniline → (HONO, $T < 8°C$) → 2,4,6-tribromobenzenediazonium → (as above) → 1,3,5-tribromobenzene

There is an alternative pattern of reactivity associated with diazonium ions. They are electrophiles and, therefore, will react with activated aromatic compounds such as phenols and phenylamines. These reactions are called coupling reactions and the products are highly-coloured dyes. For example,

a phenolic dye (red)

a phenylamine dye (orange)

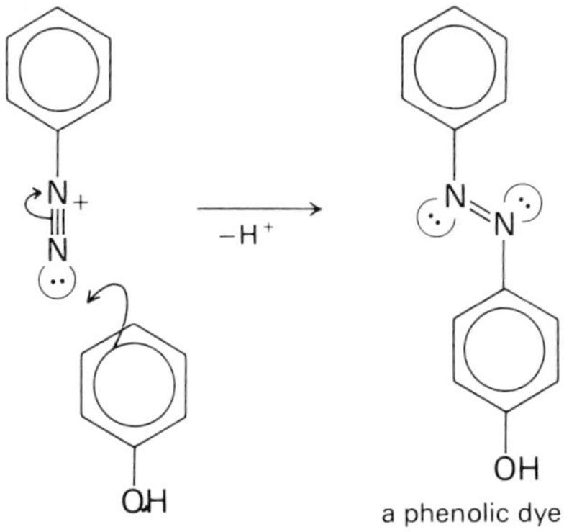

12.7 ALDEHYDES AND KETONES

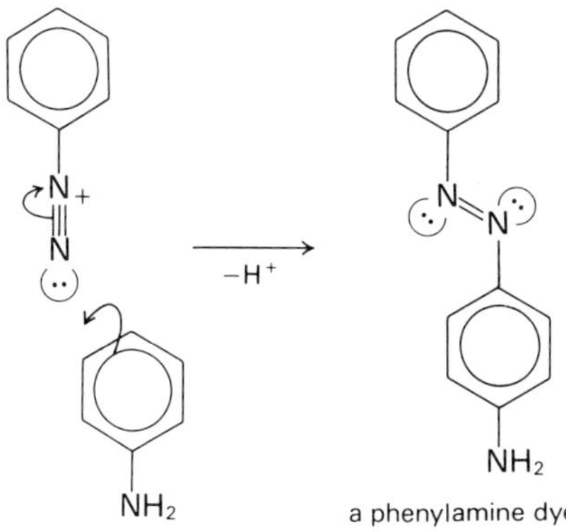

Structure, preparation and reactivity

The functional groups of a ketone and an aldehyde are shown below. Their common feature is the carbonyl group bonded to carbon or hydrogen atoms.

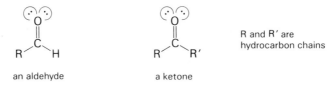

an aldehyde a ketone

R and R′ are hydrocarbon chains

The reactivity of a carbonyl group is due to the electron-deficiency of the carbon atom, the π-electrons of the carbon–oxygen bond and the lone pairs of the oxygen atom. There are two general patterns of reactivity associated with these features.

12 Classes of Organic Compounds

Firstly, a carbonyl group is particularly susceptible to nucleophilic attack. A nucleophile can be accommodated simply by the mesomeric shift of π-electrons on to the oxygen atom. After protonation, the net result if the effective addition of H—Nu.

Secondly, a carbonyl group attached to a hydrocarbon chain tends to interact with the α-hydrogen atoms of this chain. These atoms are hydrogen atoms bonded to the carbon next to the carbonyl group. A lone pair on the carbonyl oxygen atom can initiate a mesomeric shift of electrons in the molecule which results in the production of a different structural arrangement.

The alternative form of the molecule is called the *enol*- form, because it has the functional groups of an alk*ene* and an alc*ohol*. The whole effect is called *tautomerism* and is a type of structural isomerism (see page 240).

To distinguish the two tautomers from each other, the first is known as the *keto*-form. The *enol*- form reverts to the *keto*- by a cyclic shift of electrons in the opposite direction from that shown above.

The position of equilibrium usually favours the *keto*- form, but there is a sufficient proportion of the *enol*- form present to cause an alternative pattern of reactivity. As the *enol*- form is used up, the position of equilibrium adjusts to replace the amount used. The reactivity of an *enol*- is determined mostly by the donor properties of the alkene π-electrons (see page 185).

Aldehydes and ketones are usually prepared by oxidizing alcohols. This method is not very satisfactory for an aldehyde which is very readily oxidized to an acid. An alternative method for each class of compound is outlined on page 212.

aldehyde	ketone

From a primary alcohol
Make up a boiling, oxidizing solution, and drop the alcohol slowly into the vessel through a tap-funnel. A poor yield of the aldehyde is distilled off.

$$R-CH_2OH \xrightarrow[\text{in dil. acid}]{K_2Cr_2O_7} R-CHO$$

primary alcohol

From a secondary alcohol

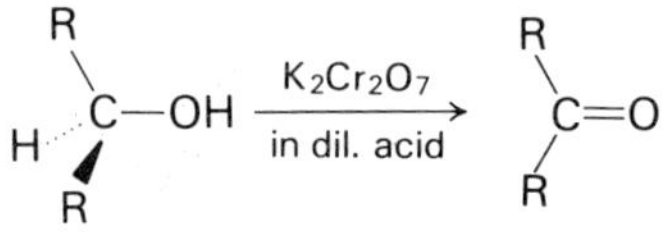

$$\underset{\text{secondary alcohol}}{\overset{R}{\underset{R}{H}}\!\!\!>\!\!C-OH} \xrightarrow[\text{in dil. acid}]{K_2Cr_2O_7} \overset{R}{\underset{R}{>}}\!\!C=O$$

From a carboxylic acid chloride (see page 223)

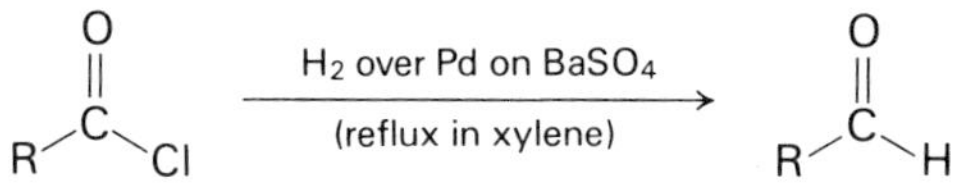

$$R-\overset{O}{\underset{}{\overset{\|}{C}}}-Cl \xrightarrow[\text{(reflux in xylene)}]{H_2 \text{ over Pd on } BaSO_4} R-\overset{O}{\underset{}{\overset{\|}{C}}}-H$$

From a carboxylic acid

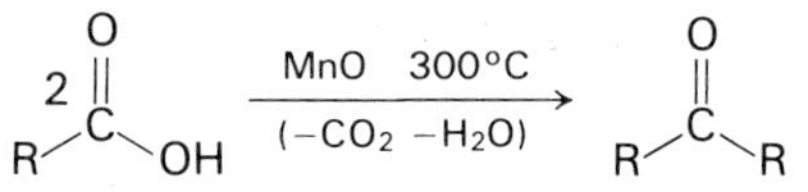

$$2\,R-\overset{O}{\underset{}{\overset{\|}{C}}}-OH \xrightarrow[(-CO_2\ -H_2O)]{MnO\ \ 300°C} R-\overset{O}{\underset{}{\overset{\|}{C}}}-R$$

Reactions

With nucleophiles. The simpler aldehydes and ketones give addition products with water. The addition is easily reversed, however, and there is probably less than 1% of the addition product present in an aqueous solution.

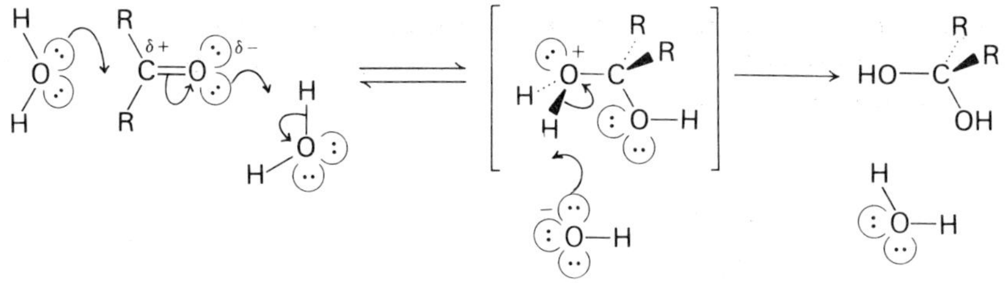

In general, a particle that contains two hydroxyl groups (or a hydroxyl group and an amino group) on the same carbon atom tends to eliminate a water molecule.

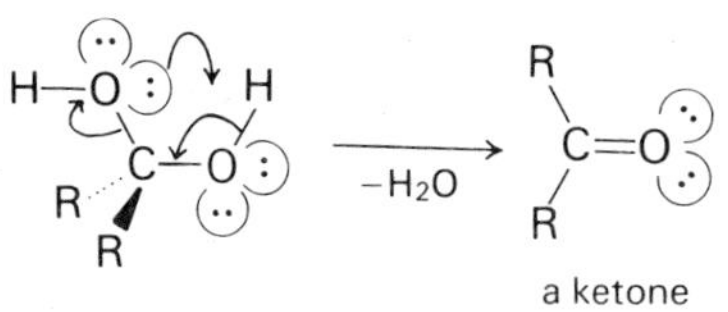

a ketone and an imine

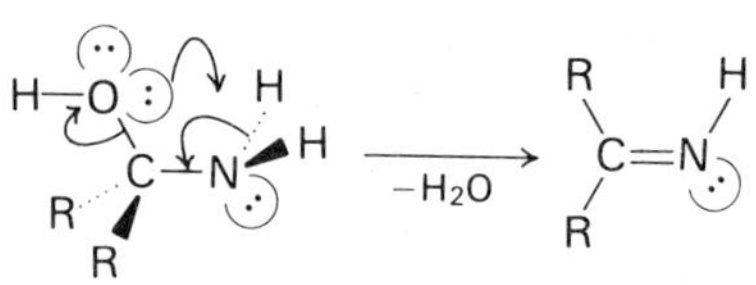

Imines are unstable and rapidly polymerize: each molecule possesses both a prominent lone pair as well as an electron-deficient carbon atom open to nucleophilic attack.

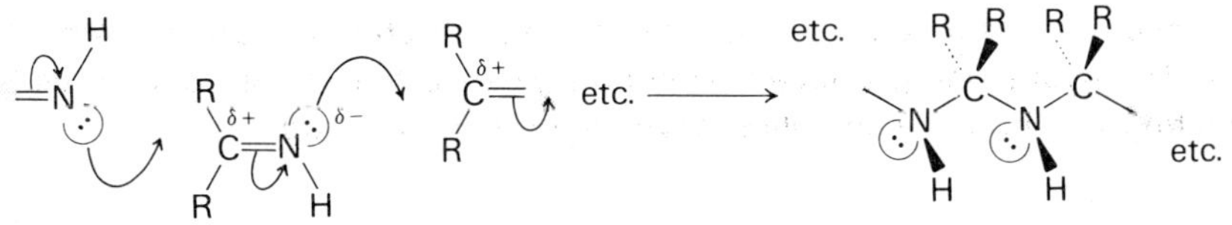

So, when ammonia is mixed with an aldehyde or ketone, the products are not simple. The initial addition product loses water to produce an imine which then polymerizes. For methanal, a cyclic trimer is formed.

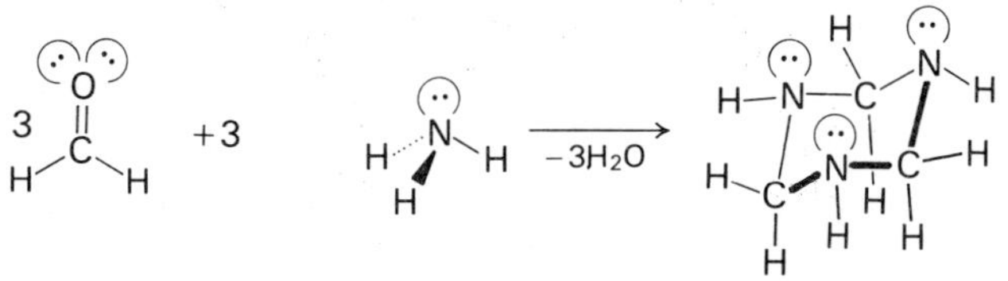

Only with the weak acid, hydrogen cyanide, does a carbonyl compound give an initial addition product that can be isolated. With hydroxylamine and hydrazine derivatives, the initial addition products eliminate water.

from the small amount of $K^+ CN^-$ present

the catalyst is regenerated

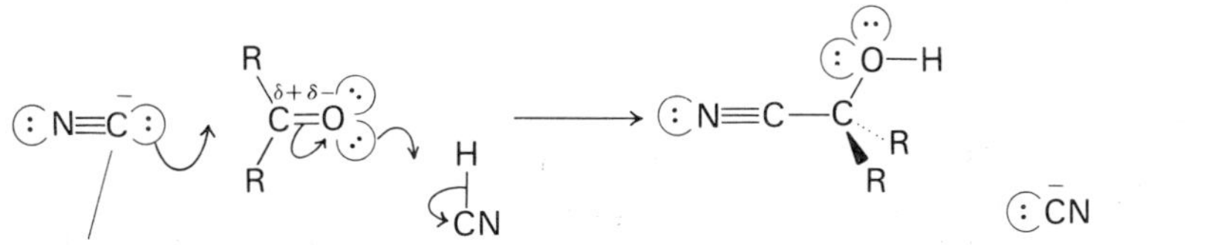

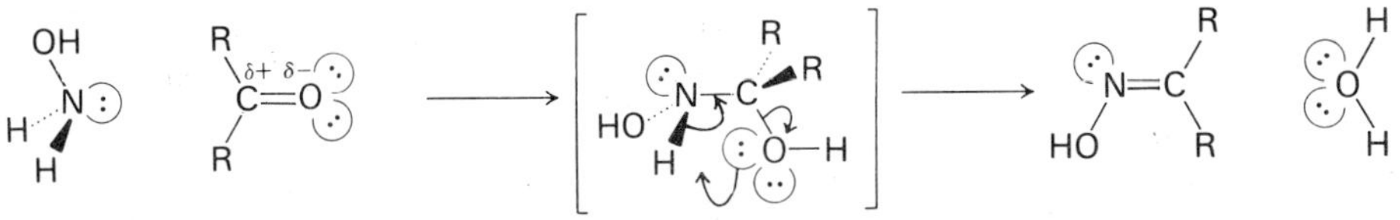

hydroxylamine | nucleophilic attack (as before) | addition product | an oxime

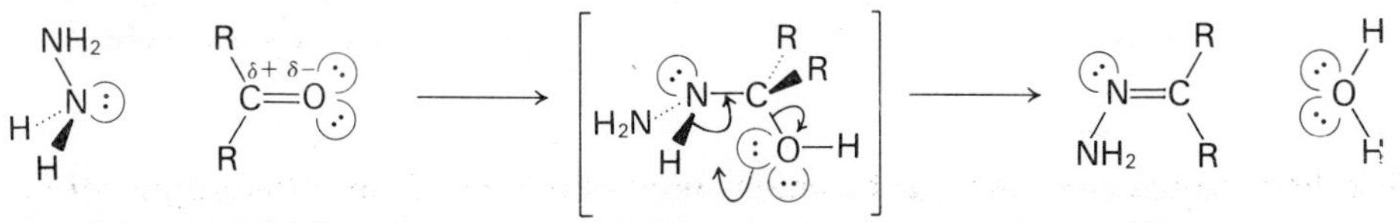

hydrazine | nucleophilic attack (as before) | addition product | a hydrazone

An oxime is produced as a crystalline precipitate when a few drops of a carbonyl compound are added to an aqueous solution of hydroxylamine. The melting point of the oxime can be used to confirm the identity of the particular carbonyl compound because tables of oxime melting points have been compiled.

For the hydrazine reaction, it is usual to use a substituted hydrazine: 2,4-dinitrophenyl-hydrazine. The hydrazone produced is an orange solid that is insoluble in water. Tables of the melting points of these solids have also been compiled.

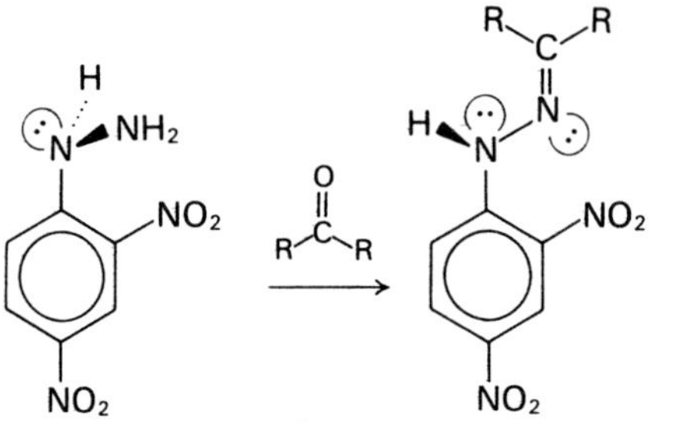

a 2,4-dinitrophenylhydrazone

With electrophiles. A carbonyl compound with α-hydrogen atoms (see page 211) is able to react with electrophiles through its *enol*- form. For example, bromine water is slowly decolorized by propanone, which has six α-hydrogen atoms per molecule.

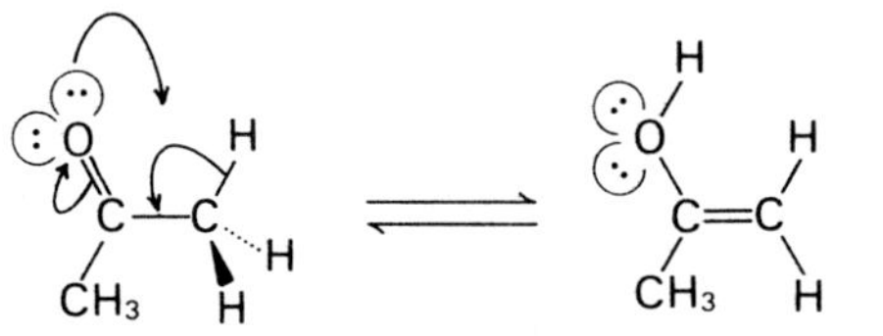

keto-propanone   *enol*-propanone

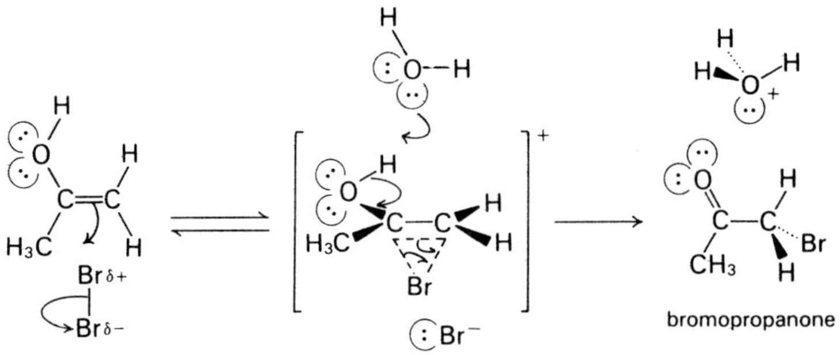

bromopropanone

The first product contains a bromine atom in place of an α-hydrogen atom. This replacement occurs for all the other α-hydrogen atoms providing there is enough bromine. The reaction is catalysed by acid which speeds up the rate of the rate-determining step: the conversion of the *keto*- to the *enol*- form. The electrophilic addition is almost instantaneous compared with the rate of the conversion. Acid assists the conversion as shown on the next page.

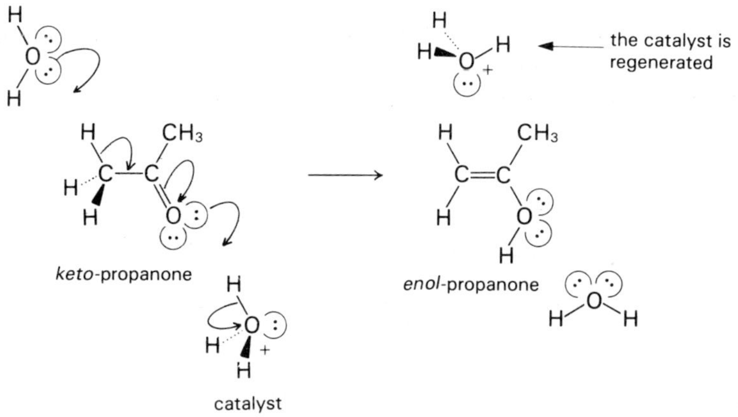

[Diagram: mechanism showing keto-propanone converting to enol-propanone via acid catalyst, with the catalyst being regenerated]

The iodoform test is a laboratory test based on the reaction of a carbonyl compound with a halogen. When a compound whose molecules contain three α-hydrogen atoms on one carbon atom is treated with iodine, the reaction shown below takes place.

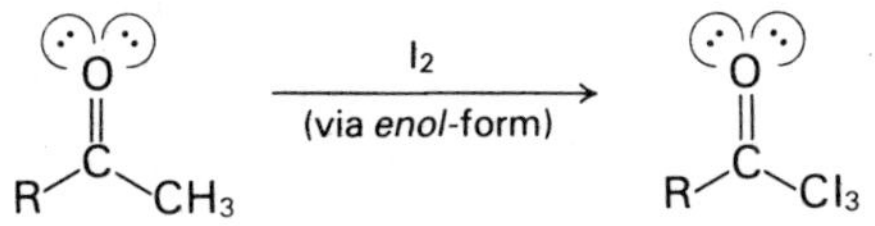

$$R-\underset{\underset{O}{\parallel}}{C}-CH_3 \xrightarrow[\text{(via enol-form)}]{I_2} R-\underset{\underset{O}{\parallel}}{C}-CI_3$$

The iodoform reagent is made up under alkaline conditions and the triodo-product is readily hydrolysed to iodoform.

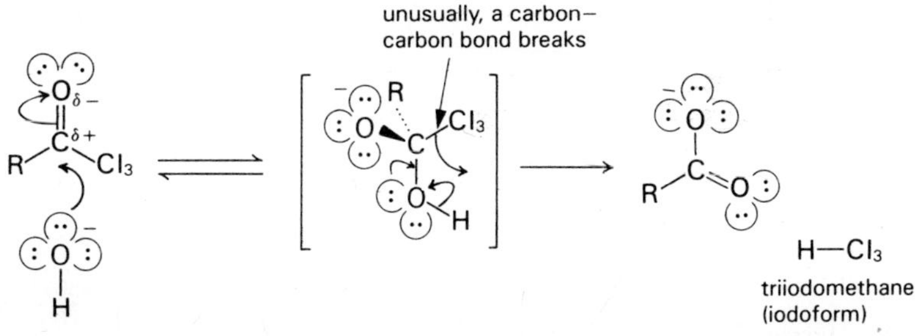

[Diagram: mechanism showing hydroxide attack on $R-CO-CI_3$, unusually a carbon–carbon bond breaks, producing $R-COO^-$ and $H-CI_3$ (triiodomethane, iodoform)]

The iodoform produced is an insoluble yellow solid, whose appearance provides confirmation of the identity of the original compound tested. It must have either of the structures shown below.

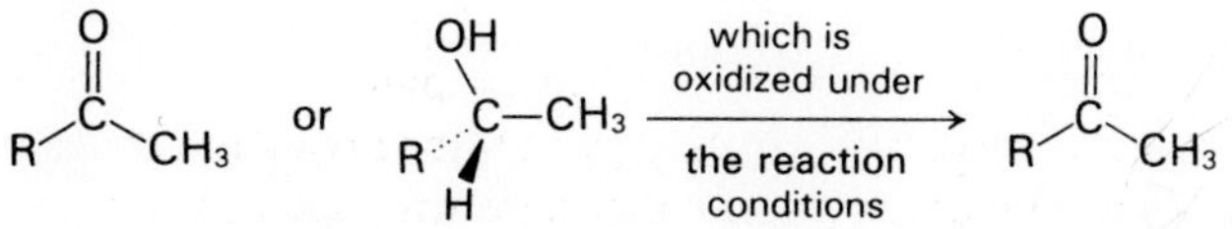

$$R-\underset{\underset{O}{\parallel}}{C}-CH_3 \quad \text{or} \quad R-\underset{\underset{H}{|}}{\overset{OH}{\underset{|}{C}}}-CH_3 \xrightarrow[\text{the reaction conditions}]{\text{which is oxidized under}} R-\underset{\underset{O}{\parallel}}{C}-CH_3$$

For example, the following compounds give positive results when a few drops are added to a solution of iodine in potassium iodide containing a little alkali:

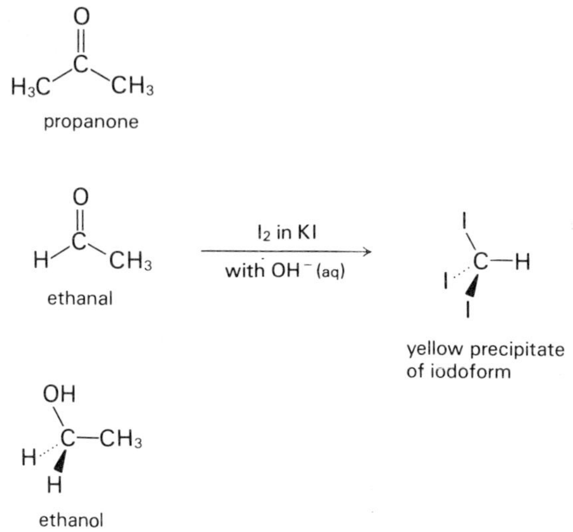

propanone: H₃C-CO-CH₃

ethanal + I₂ in KI with OH⁻(aq) → CHI₃ (yellow precipitate of iodoform)

ethanol: CH₃-CH(OH)-H

With redox reagents. To distinguish between an aldehyde and a ketone, it is possible to make use of the fact that an aldehyde can be oxidized to an acid, but that a ketone is resistant to oxidation. Aldehydes can be oxidized by silver(I) ions when complexed by ammonia molecules. Silver(0) is one of the products, and so a 'silver mirror' forms when an aldehyde is warmed with ammoniacal silver nitrate. This is called the silver mirror test; ketones do not give positive results.

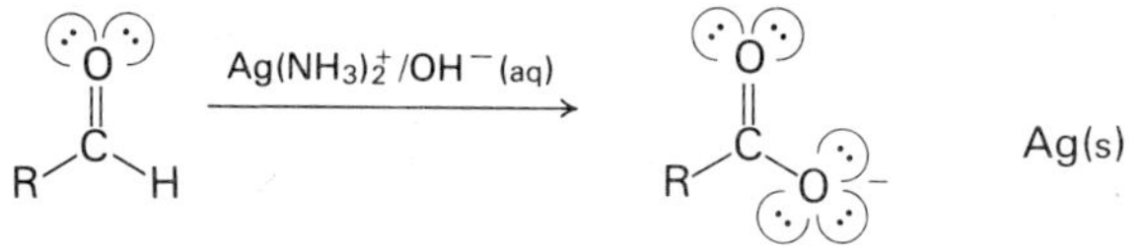

$$RCHO \xrightarrow{Ag(NH_3)_2^+ / OH^- (aq)} RCOO^- + Ag(s)$$

Both aldehydes and ketones are reduced by lithium tetrahydridoaluminate, which is a source of hydride ions (see page 115).

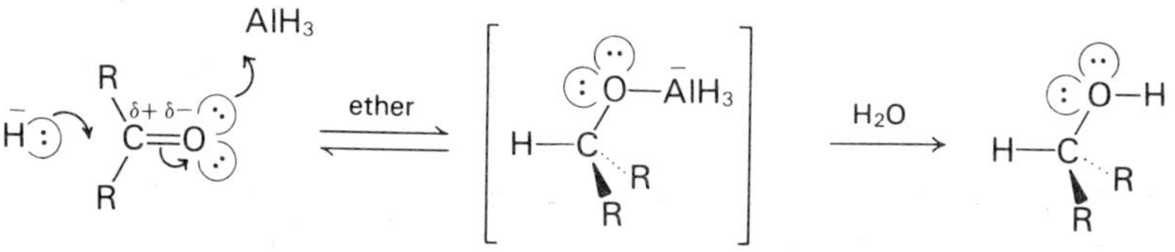

$$R_2C=O + H^- \xrightarrow[\text{ether}]{AlH_3} [R_2CH-O-AlH_3] \xrightarrow{H_2O} R_2CH-OH$$

They are also reduced by Raney nickel in ethanolic solution. Raney nickel is very finely divided nickel containing adsorbed hydrogen. It is prepared by dissolving the aluminium from a nickel–aluminium alloy in a solution of sodium hydroxide.

$$2Ni\text{-}Al(s) + 2OH^-(aq) + 6H_2O(l) \longrightarrow 2Al(OH)_4^-(aq) + 2Ni\text{-}3H_2(s)$$

Raney nickel

$$R_2C=O \xrightarrow[\text{or 1 LiAlH}_4 \; 2\,H_2O]{\text{either Raney nickel}} R_2CH-OH$$

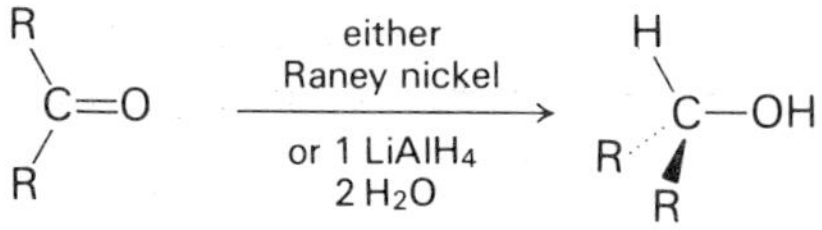

Base-catalysed condensation. A small concentration of carbanions is generated by deprotonating the *enol-* form of a carbonyl compound. These carbanions are powerful nucleophiles and attack the *keto-* form of the carbonyl compound to produce condensation products. A typical case is the aldol reaction: ethanal in very dilute alkali gives the dimer, 3-hydroxybutanal (also known as aldol).

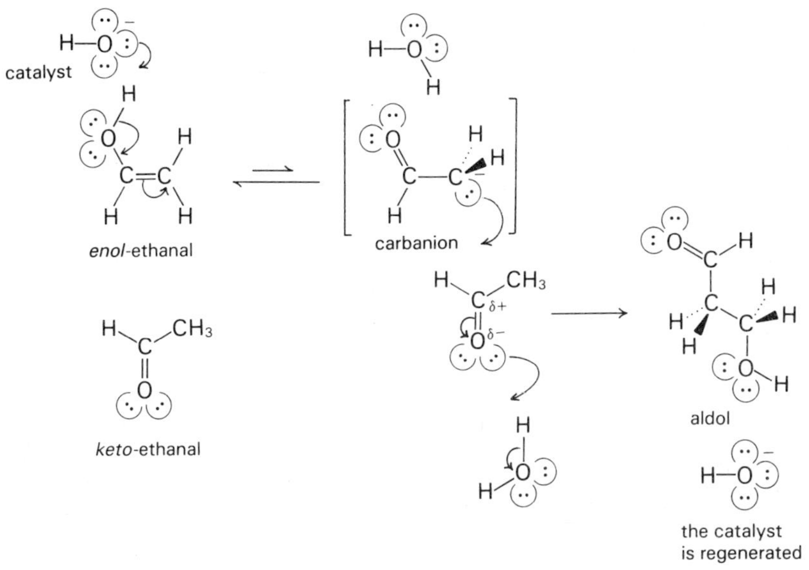

In more concentrated alkali, the condensation process continues further until a polymeric brown resin is produced. The structure of the resin contains some unsaturation due to dehydration. For example,

polymeric resin

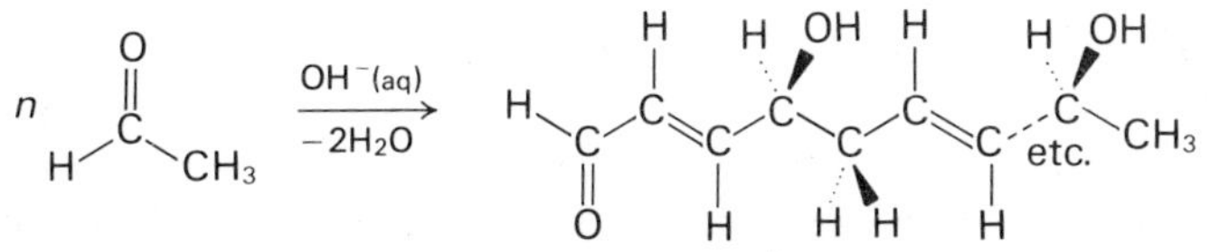

Other carbonyl compounds condense under more strongly basic conditions. For example, even ethyl ethanoate condenses when refluxed with sodium ethoxide (the Claisen condensation).

ethyl 3-oxybutanoate
(a β-keto ester)

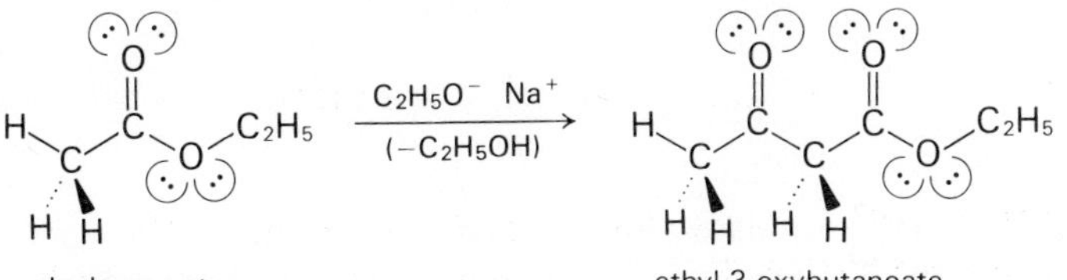

12.8 CARBOXYLIC ACIDS

Structure, preparation and reactivity

The functional group of a carboxylic acid is a carbonyl group bonded directly on to a hydroxyl group.

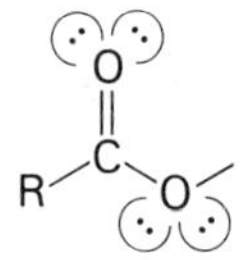

The properties of the two groups are modified by their close proximity. Conjugation takes place between the hydroxyl lone pair and the carbonyl π-system.

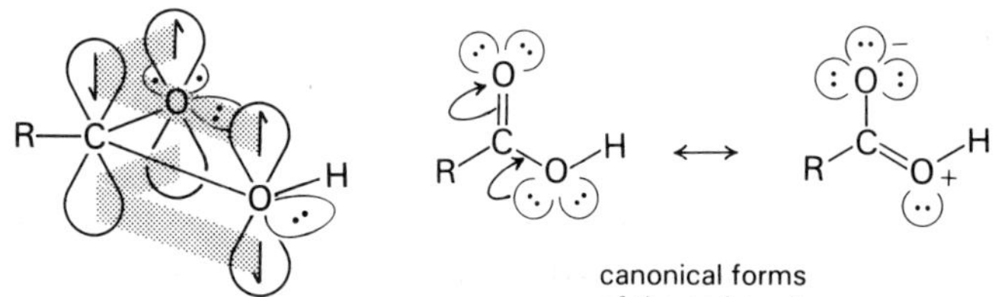

A

B ↔ *C*

canonical forms of the carboxylic acid structure (see page 67)

Both the molecular orbital (*A*) and valence bond models (*B* and *C*) shown above, illustrate the position of the hydrogen atom in the hydroxyl group: it is held in a plane between the two oxygen atoms. The dominant feature of this structure is the ease with which it loses a proton (the nucleus of the hydrogen atom). The charge of the anion produced is stabilized by the conjugation effect. In particular, it can be seen that the canonical form *C* is far more stable as a result of the proton loss:

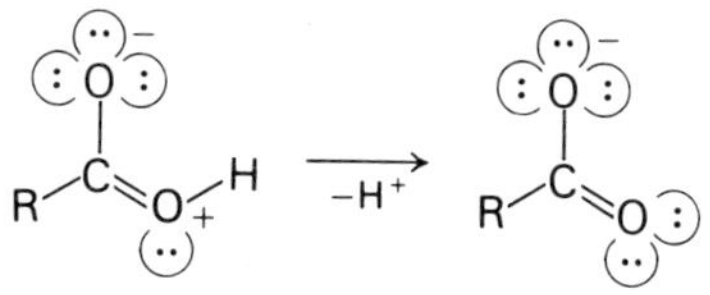

The nature of R— has a marked effect on the acid strength. If R— exerts a $-I$ or $-M$ effect, the acid strength increases appreciably because the charge of the anion is delocalized on R. For example, consider the strengths of the acids shown on the next page.

identity of R— in R—COOH	electron-releasing or electron-withdrawing effect of R— compared with hydrogen	pK_a
H—	standard	3.75
CH_3—	releasing (+I)	4.76
CCl_3—	withdrawing (−I)	0.65
2-aminomethylphenyl (with NH_2)	releasing (+M)	6.97
2-nitromethylphenyl (with NO_2)	withdrawing (−M)	2.17

The reactivity expected of the carbonyl group is considerably reduced when it is conjugated to a hydroxyl group. Nucleophilic attack is less likely for two reasons: the nucleophile is often protonated instead, and even if the nucleophile does approach the electron-deficient carbon atom, it is more difficult to displace the π-electrons from the conjugated π-system.

Similarly, the reactivity of the hydroxyl group in an acid molecule is altered because of the conjugation. Nucleophilic properties are almost totally absent, and the ability of the group to act as a leaving-group is also restricted.

An acid is prepared by oxidizing a primary alcohol or an aldehyde, or by the hydrolysis of a nitrile. Acid derivatives (such as esters, amides, acid chlorides and anhydrides) are also readily hydrolysed to their parent acid.

$$R-CH_2OH \text{ or } R-CHO \xrightarrow{K_2Cr_2O_7 \text{ in dilute acid}} R-COOH$$

$$R-C\equiv N \xrightarrow[\text{or 1. } OH^-\;\; 2. H_3O^+]{\text{either 1. } H_3O^+\;\; 2. OH^-} R-COOH$$

Reactions

With bases. Soluble acids form salts with inorganic bases. They are neutralized by metal oxides, hydroxides and carbonates in the same way that inorganic acids are. For example,

$$2CH_3COOH + PbO \longrightarrow (CH_3COO)_2Pb + H_2O$$
$$2CH_3COOH + Na_2CO_3 \longrightarrow 2CH_3COONa + H_2O + CO_2$$

Many carboxylic acids are insoluble in water, however. Their hydrocarbon chains are non-polar and have little attraction for polar water molecules. Even when soluble, the acids are weak and exist mostly as molecules in solution. All acids are far more soluble in alkali.

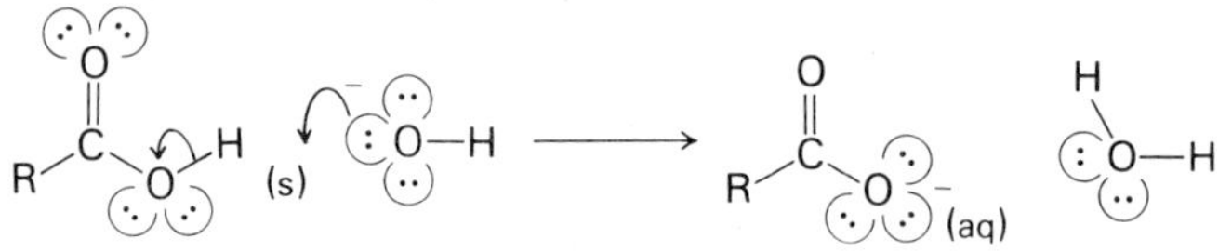

Precipitation occurs when a mineral acid is added to a solution of the sodium salt of an insoluble carboxylic acid.

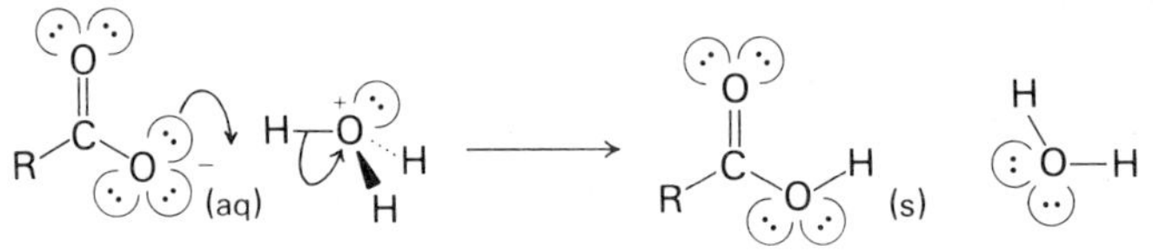

An acid also neutralizes aqueous ammonia to give an ammonium salt. Dehydration of this salt produces first an amide and subsequently a nitrile.

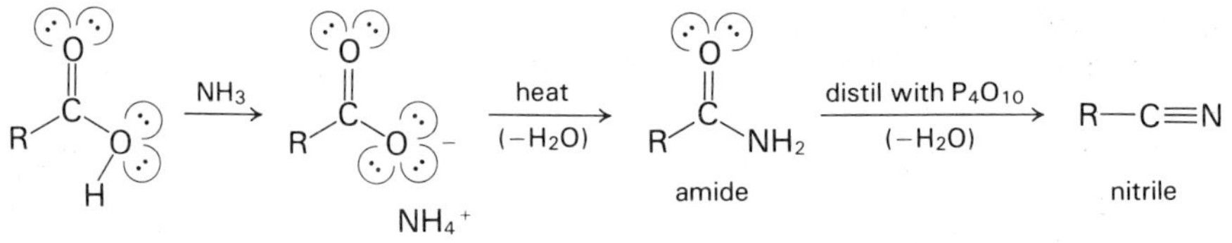

With nucleophiles. Hydroxide ions and ammonia molecules are protonated by carboxylic acids and so cannot act as nucleophiles. An alcohol, however, reacts with a carboxylic acid in the presence of concentrated sulphuric acid to produce an ester. The concentrated sulphuric acid makes the hydroxyl group more likely to act as a leaving group by protonating it and converting it to a water molecule. Protonated carboxylic acid molecules are far more open to nucleophilic attack (compare with the discussion on page 200).

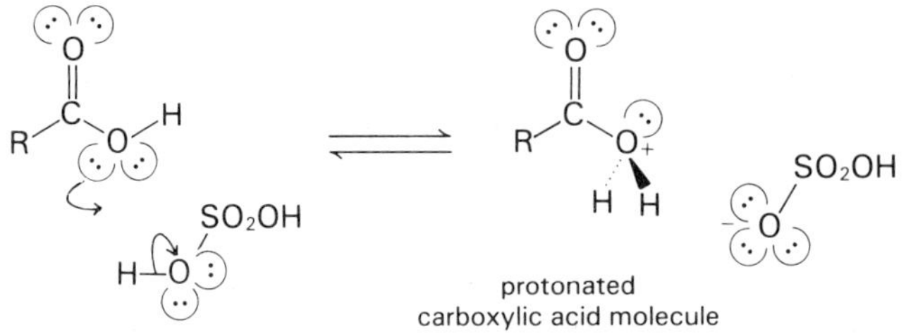

A protonated acid molecule is attacked by an alcohol molecule, and a water molecule leaves to produce an ester molecule.

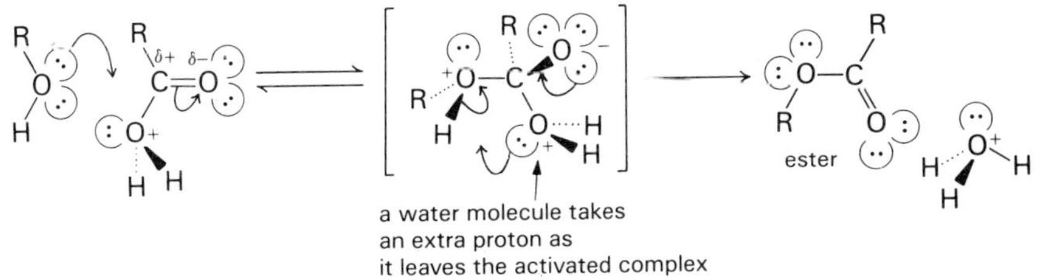

a water molecule takes an extra proton as it leaves the activated complex

One acid molecule acts as a nucleophile in attacking another acid molecule when an acid is distilled in the presence of the strongly acidic oxide, phosphorus pentoxide. Water is eliminated and an acid anhydride is produced.

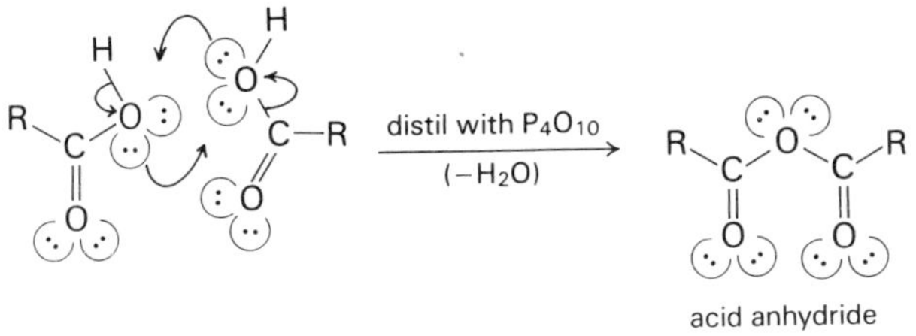

acid anhydride

Esters and anhydrides are best prepared from acid chlorides, however. An acid is easily converted to its acid chloride by adding phosphorus pentachloride.

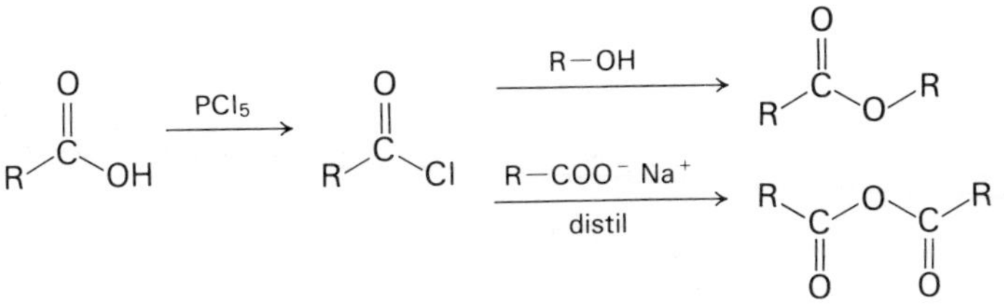

With reductants. The only reductants capable of reducing a carboxylic acid are those that act as hydride donors. The reactions are nucleophilic processes involving hydride ions as nucleophiles. Lithium tetrahydridoaluminate (page 216) or a borane can be used, and the product is a primary alcohol.

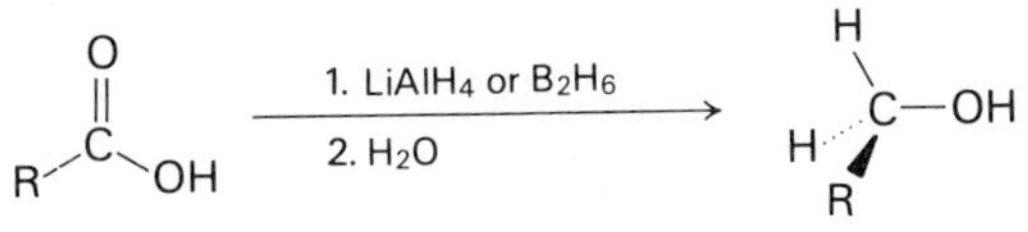

12.9 ACID CHLORIDES AND ANHYDRIDES

Structure, preparation and reactivity

Acid chlorides and anhydrides are carbonyl compounds whose molecules contain a carbonyl group bonded to a likely leaving group. In an acid chloride, the leaving group is a chloride ion; in an acid anhydride, the leaving group is a carboxylate ion.

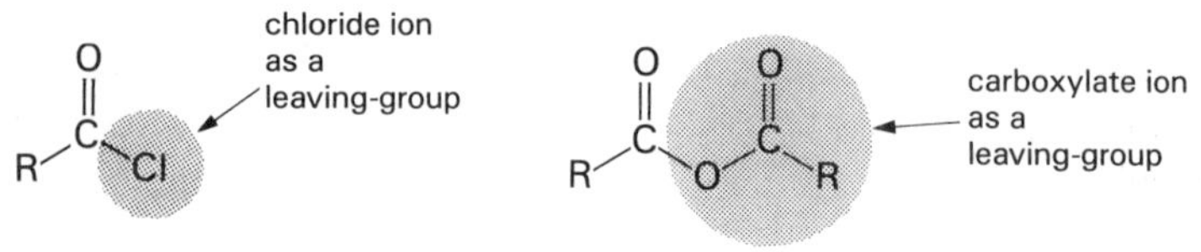

These 'acid derivatives' are attacked very vigorously by nucleophiles. The attacking nucleophile is easily accommodated by a mesomeric shift of the carbonyl π-electrons, and the subsequent activated complex is rapidly stabilized by the loss of the leaving group.

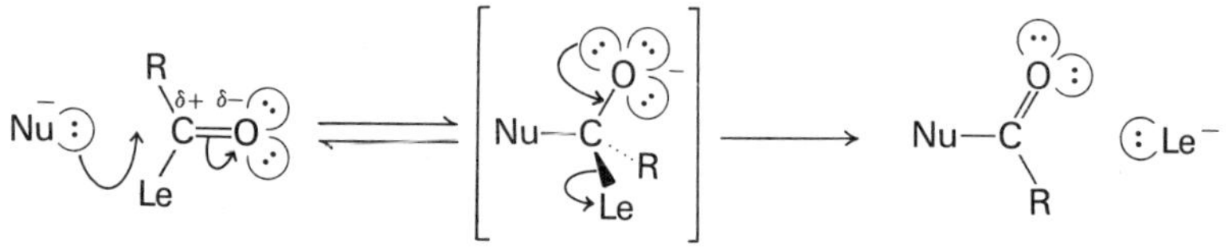

An acid chloride is prepared by treating an acid with phosphorus pentachloride or sulphurdichloride oxide. An acid anhydride is best prepared by distilling a mixture of an acid chloride and the sodium salt of an acid. However, the dehydration of an acid also produces an anhydride as described earlier.

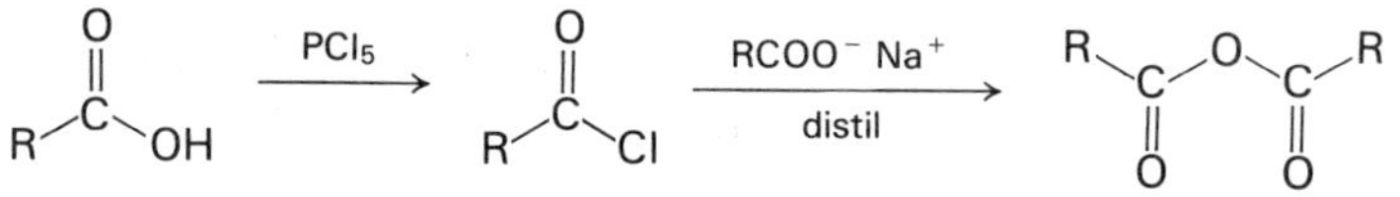

Reactions

With nucleophiles. Acid chlorides and anhydrides react with even the weakest of nucleophiles such as water and aromatic amines. Some examples of these reactions are given below. Note that the preparation of an anhydride from an acid chloride and salt also follows this pattern; the carboxylate ion acts as a nucleophile.

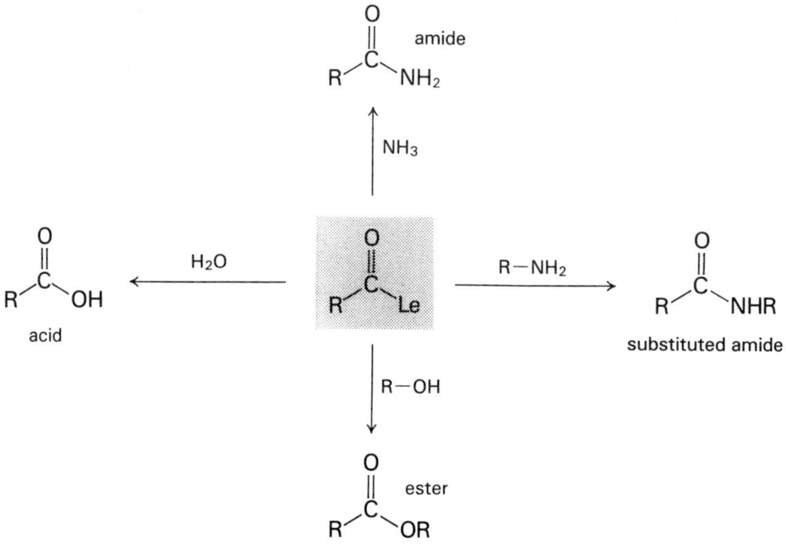

When ethanoyl chloride (or anhydride) reacts with a nucleophilic reagent, the process is known as ethanoylation. Ethanoylation results in the substitution of a hydrogen atom in the nucleophile by the ethanoyl group.

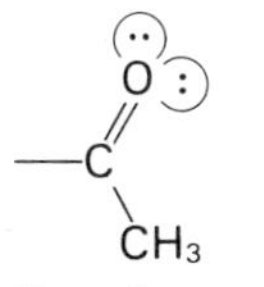

ethanoyl group

For example, an alcohol is ethanoylated to an ethanoate ester and an amine to a substituted ethanamide.

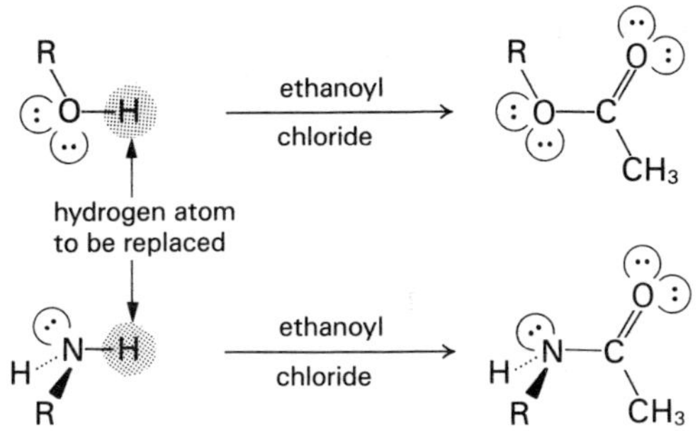

Even benzene can be ethanoylated, providing there is some aluminium chloride present. This is the Friedel–Crafts reaction (see page 187).

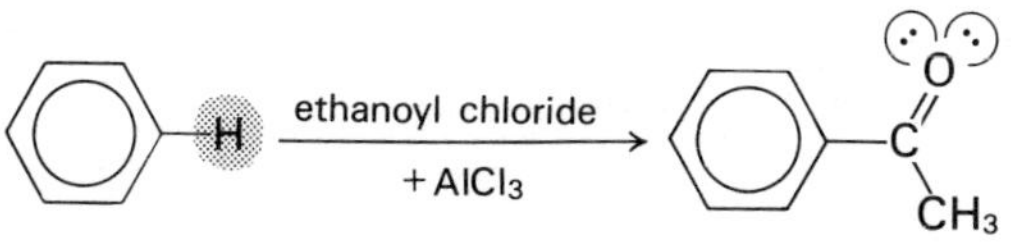

Similarly, benzoylation involves the reactions of benzoyl chloride or anhydride.

With reductants. Like all acids and their derivatives, acid chlorides and anhydrides are reduced to alcohols by lithium tetrahydridoaluminate. More useful from the point of view of synthesis, is the reduction of an acid chloride by hydrogen in boiling dimethylbenzene (xylene). A catalyst of finely divided palladium on barium carbonate is used and the product is an aldehyde. It is difficult to prepare an aldehyde because of the ease with which the aldehyde itself undergoes redox reactions. Using this method (Rosenmund reduction) a reasonable yield can be achieved.

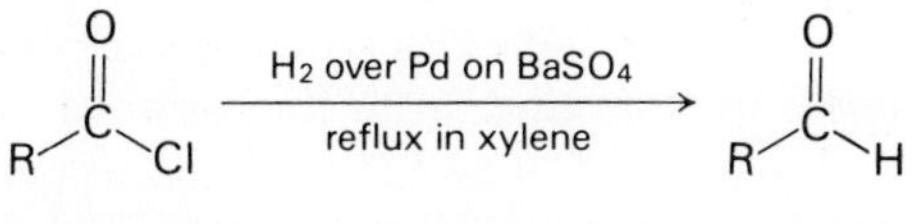

12.10 ESTERS AND AMIDES

Structure, preparation and reactivity

The functional group of an ester or amide takes the form of a particular type of molecular link.

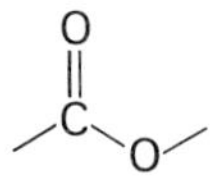

ester link

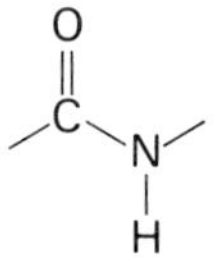

amide or peptide link

As in the case of acids, the carbonyl group is considerably affected by conjugation. For example, methyl ethanoate and ethanamide are shown below.

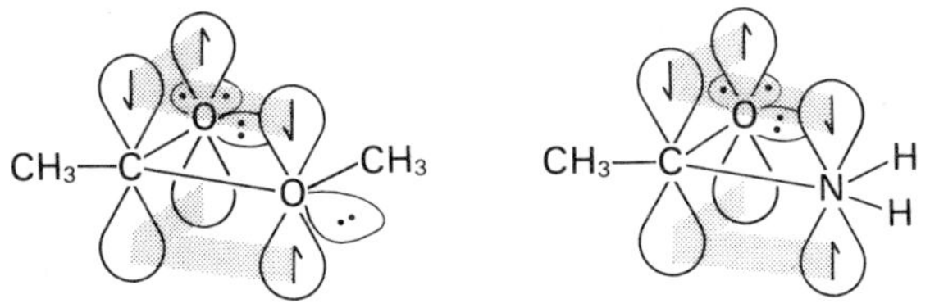

Esters and amides have a greater affinity for nucleophiles than acids have. An attacking nucleophile is often protonated by an acid, but this cannot happen in the case of an ester or an amide. However, as a result of the conjugation, the carbonyl group is still less reactive than the carbonyl group of a ketone. For example, amides and esters do not give precipitates with hydroxylamine or 2,4-dinitrophenylhydrazine.

The structure of an unsubstituted amide (such as ethanamide above) shares a common feature with that of an acid shown on page 218. A proton is trapped in the plane of the carbonyl group as a result of the attraction to an oxygen lone pair. The environment of the second hydrogen atom can be seen to be very different. It is, in fact, possible for an amide itself to act as an acid in losing this proton; for example, with mercury(II) oxide.

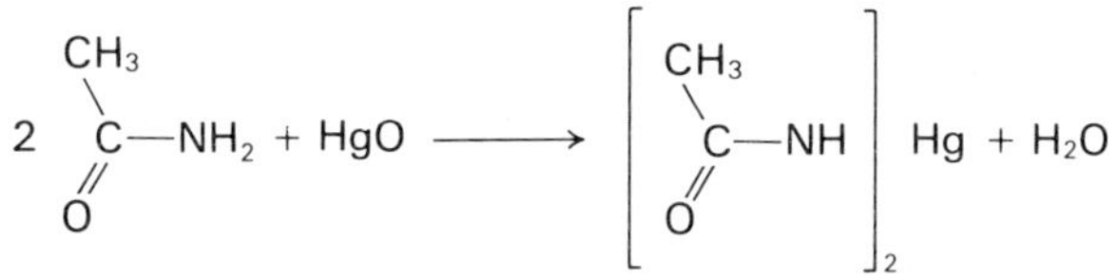

$$2\ CH_3-\underset{\underset{O}{\|}}{C}-NH_2 + HgO \longrightarrow [CH_3-\underset{\underset{O}{\|}}{C}-NH]_2 Hg + H_2O$$

acid   base   salt   water

This behaviour contrasts sharply with the basic properties usually associated with an amino-group. Amides do *not* act as bases or nucleophiles because the nitrogen lone pair is extensively delocalized by conjugation to the carbonyl π-system.

12 Classes of Organic Compounds

Esters and amides are prepared by reacting an alcohol or an amine with an acid derivative.

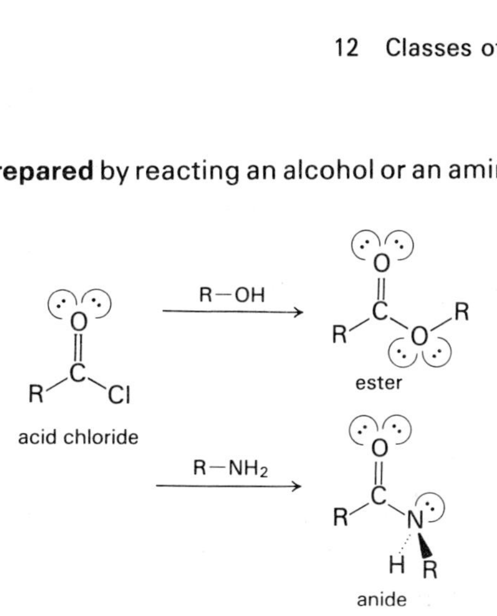

acid chloride → ester (with R–OH)

acid chloride → amide (with R–NH₂)

A variety of macromolecular compounds contain ester or amide (peptide) links. Naturally occurring polyamides are known simply as peptides (or polypeptides), and these include the important class of compounds called proteins. Each protein is derived from a large number of α-aminoacid molecules which have undergone condensation by forming peptide links as shown below (an α-aminoacid contains an amino-group bonded to the carbon atom next to the carboxylic acid group).

(i) two α-aminoacid molecules → peptide link forged ($-H_2O$)

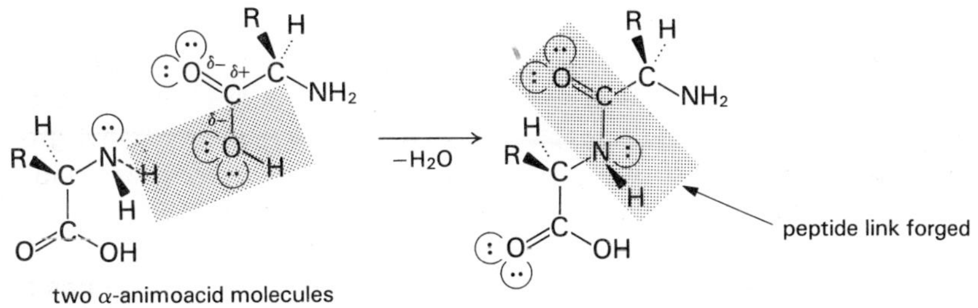

The structure of a typical protein contains a sequence of different α-aminoacids characteristic of the particular protein.

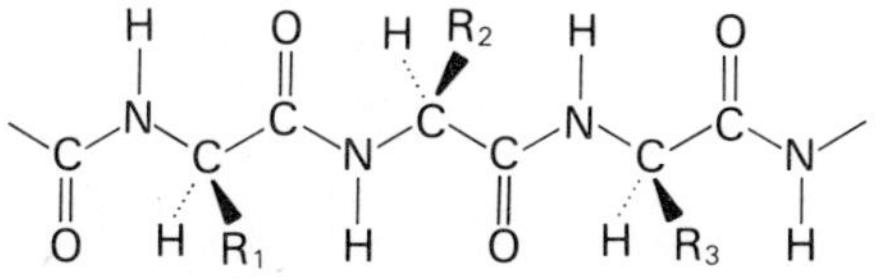

a typical protein chain

R_1 R_2 R_3 etc are different hydrocarbon chains

Synthetic polyamides are produced by condensing a diamine with a diacid (see page 177). The macromolecular chains obtained from this process are used as fibres and are given the general name of nylons. Synthetic polyesters are manufactured in a similar way by reacting a diol with a diacid (or sometimes a diester as on page 172).

The ester link is also found in naturally occurring substances, and these include fats and the 'nucleic' acids present in the nuclei of all living cells. Fats are esters based on the triol, glycerol. Long chain acids are obtained by the hydrolysis of a fat, and so these carboxylic acids are often called fatty acids.

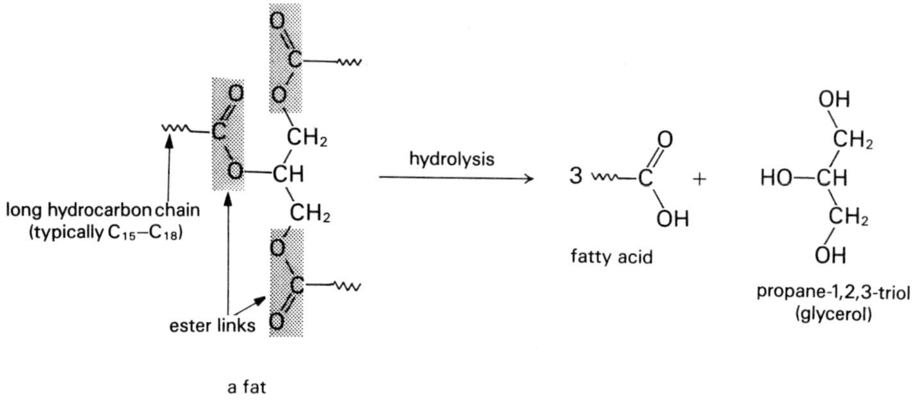

Nucleic acids contain phosphate ester links instead of carboxylate ester links. A small part of the chain of a deoxyribonucleic acid (DNA) is shown below.

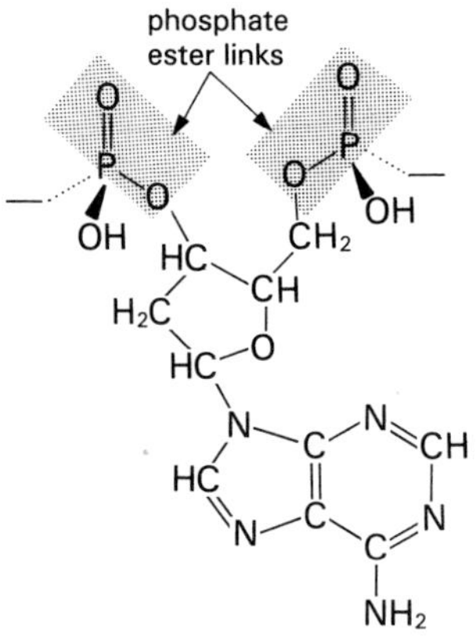

which follows the general pattern shown on the right

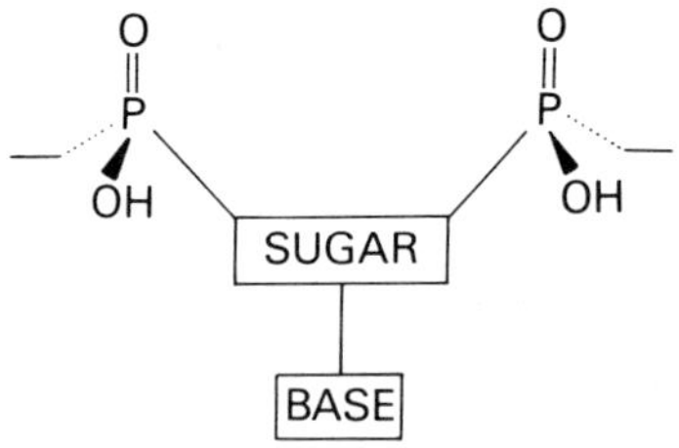

Reactions

With nucleophiles. Esters and amides are hydrolysed under either alkaline or acidic conditions. In the first case, hydroxide ions act as nucleophiles; in the second case, a water molecule attacks the carbonyl group as a result of the simultaneous protonation of the oxygen atom as shown on the next page.

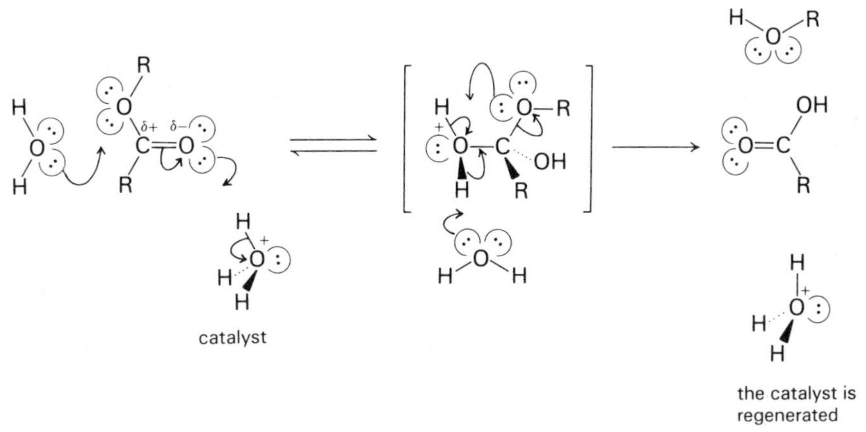

A laboratory test for an unsubstituted amide derives from the ability of an amide to produce ammonia when boiled in alkali. The need for heat distinguishes an amide from an ammonium salt which evolves ammonia in the cold.

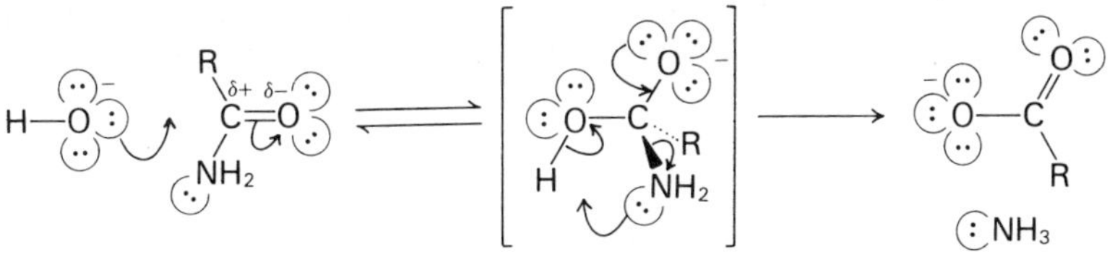

Nitriles behave in the same way as esters and amides under alkaline or acidic conditions. A nitrile is first hydrolysed to an amide which is then subsequently hydrolysed further. For example, the alkaline hydrolysis is shown below, and the acidic hydrolysis is shown over the page.

in alkali:

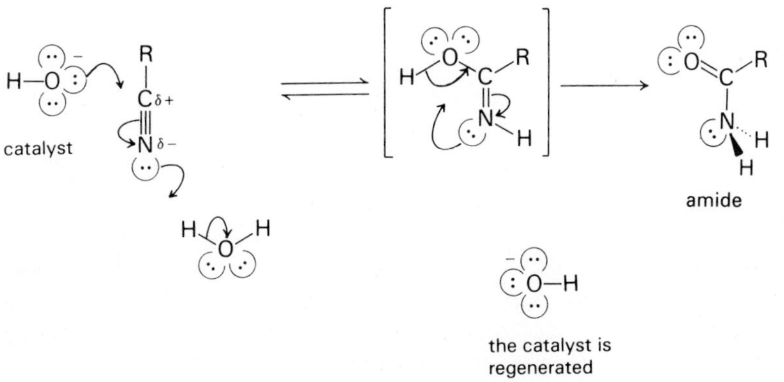

in acid:

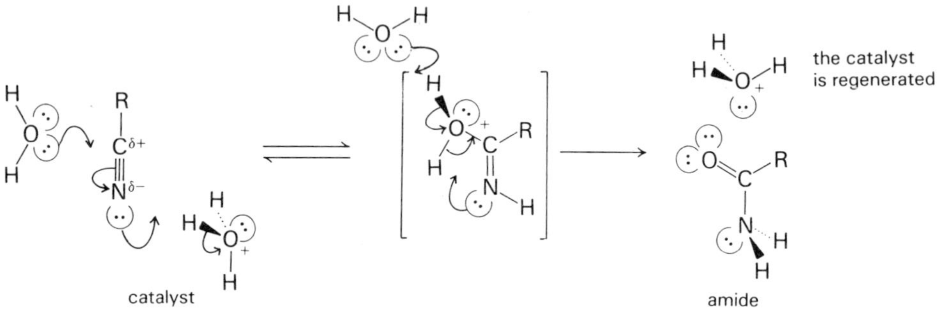

[mechanism diagram showing acid-catalyzed formation of amide from intermediate, with labels "catalyst", "amide", and "the catalyst is regenerated"]

An ester can be converted to an amide by reaction with concentrated aqueous ammonia. The ammonia molecules are strong enough nucleophiles for reaction to occur.

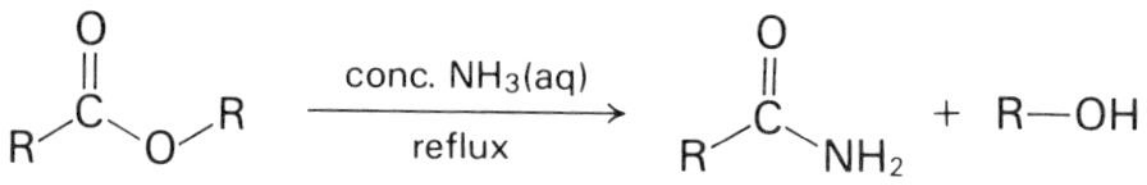

$$R-\underset{\underset{O}{\|}}{C}-O-R \xrightarrow[\text{reflux}]{\text{conc. NH}_3\text{(aq)}} R-\underset{\underset{O}{\|}}{C}-NH_2 + R-OH$$

With reductants. Lithium tetrahydridoaluminate reduces an ester to an alcohol, and an amide or nitrile to an amine. The mechanism by which this reductant functions is described on page 216.

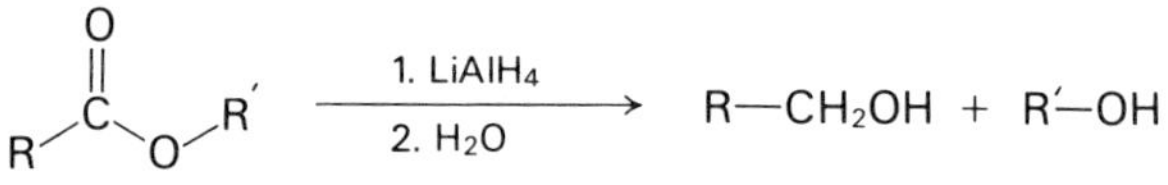

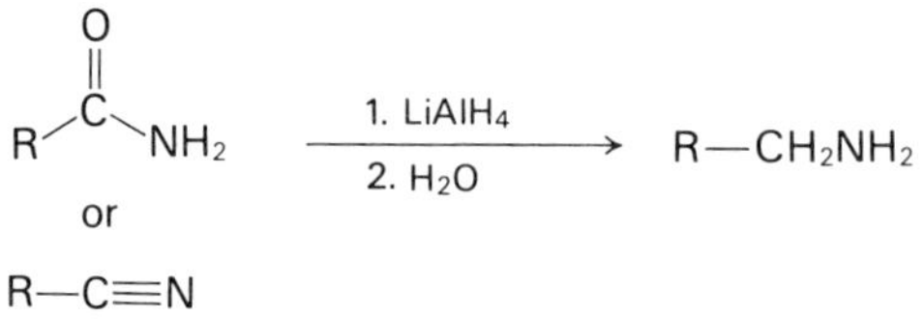

$$R-\underset{\underset{O}{\|}}{C}-O-R' \xrightarrow[\text{2. H}_2\text{O}]{\text{1. LiAlH}_4} R-CH_2OH + R'-OH$$

$$R-\underset{\underset{O}{\|}}{C}-NH_2 \xrightarrow[\text{2. H}_2\text{O}]{\text{1. LiAlH}_4} R-CH_2NH_2$$

or

$$R-C\equiv N$$

Nitriles can also be reduced catalytically, but ammonia must be added to the reaction mixture to prevent the formation of secondary amines. In the absence of ammonia, the amine produced attacks the imino-intermediate which represents the half-reduced stage; in the presence of ammonia the ammonia molecules act as the nucleophiles instead, and yields of the primary amine are then much higher.

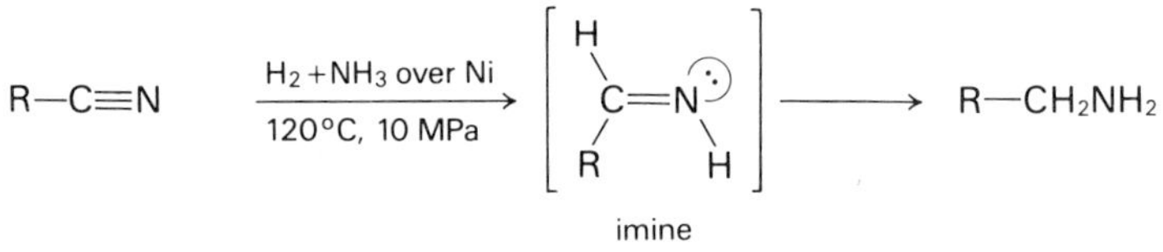

$$R-C\equiv N \xrightarrow[120°C,\ 10\ \text{MPa}]{H_2+NH_3\ \text{over Ni}} \left[\underset{R}{\overset{H}{\diagdown}}C=N\underset{H}{\diagdown} \right] \longrightarrow R-CH_2NH_2$$

imine

12 Classes of Organic Compounds

As ligands. A polypeptide that contains two peptide links conjugated together, gives a characteristic complex with copper(II). Polypeptides of this sort can have one of the following structures:

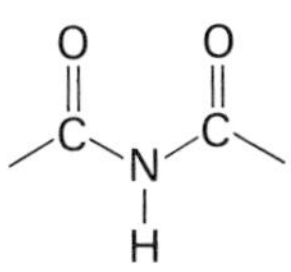

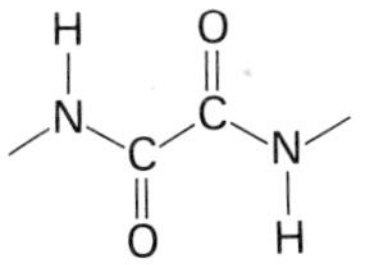

A pale pink coloured complex is obtained when the peptide is added to very dilute copper(II) sulphate solution containing a trace of alkali. The test is often called the biuret test after the condensation product of urea which satisfies the structural requirements.

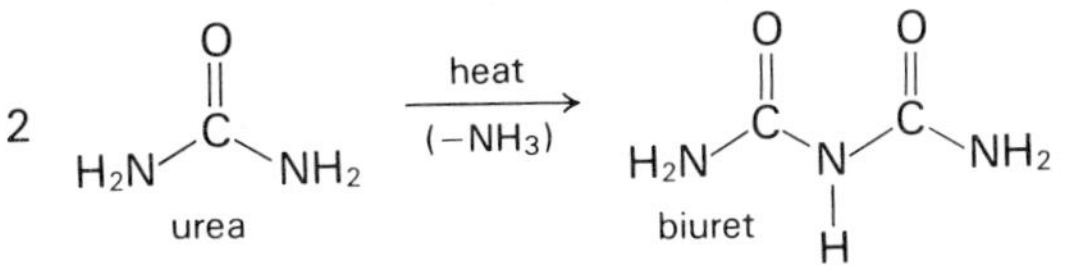

$$2 \; H_2N{-}CO{-}NH_2 \xrightarrow[(-NH_3)]{\text{heat}} H_2N{-}CO{-}NH{-}CO{-}NH_2$$

urea → biuret

Ethanediamide is another common compound that gives positive results.

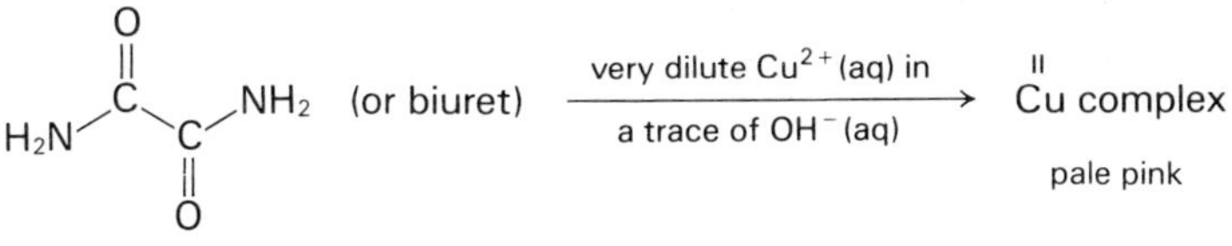

$$H_2N{-}CO{-}CO{-}NH_2 \;(\text{or biuret}) \xrightarrow[\text{a trace of } OH^-(aq)]{\text{very dilute } Cu^{2+}(aq) \text{ in}} \text{Cu complex (pale pink)}$$

ethanediamide

The Hofmann degradation reaction. The hydrogen atoms of the amino-group in an unsubstituted amide are α-hydrogen atoms. There is, therefore, a slight tendency for an amide to exist in an *enol-* form.

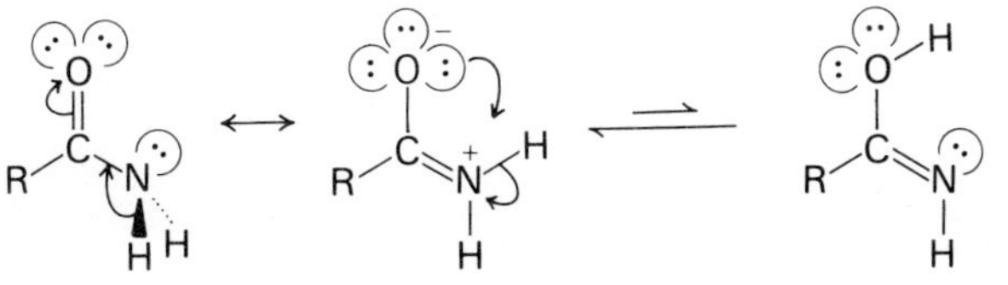

canonical forms of *keto*-amide ⇌ *enol*-amide

The *enol-* tautomer reacts with electrophiles such as bromine. As a result of reacting with bromine, an α-hydrogen atom is replaced by a bromine atom (see page 214).

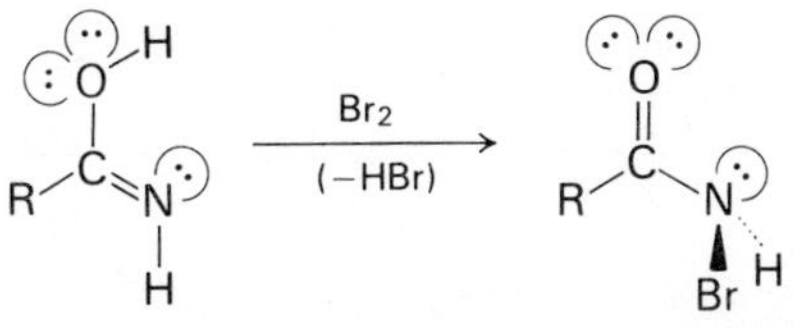

$$R{-}C(OH){=}N{-}H \xrightarrow[(-HBr)]{Br_2} R{-}CO{-}N(Br){H}$$

If this reaction is carried out in the presence of alkali, the bromo-derivative is fairly readily deprotonated and the anion produced rapidly rearranges by eliminating a bromide ion.

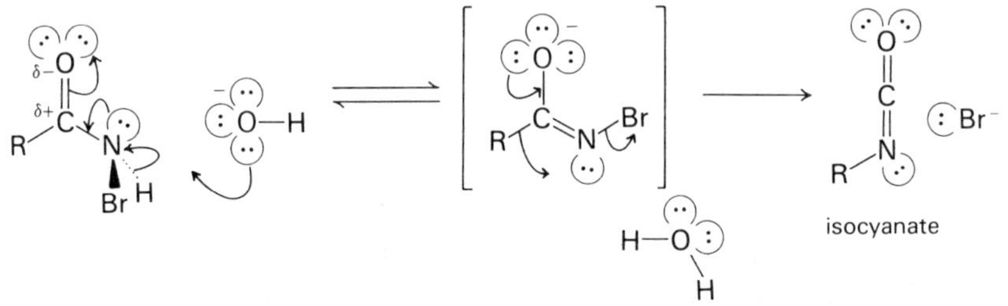

The isocyanate produced is unstable in alkali and is immediately hydrolysed to an amine and carbon dioxide.

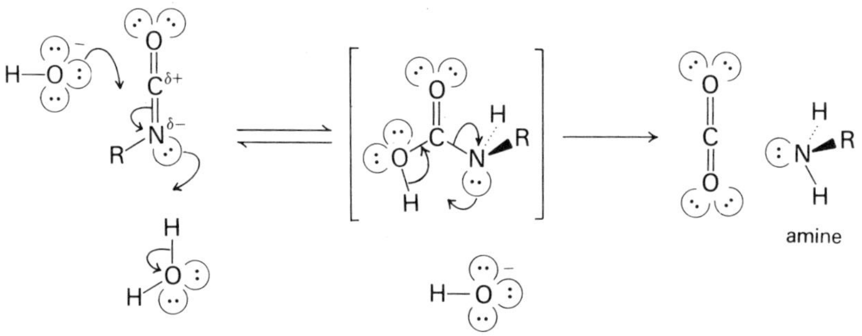

This overall process is called the Hofmann degradation because it achieves the removal of one carbon atom from the molecular chain.

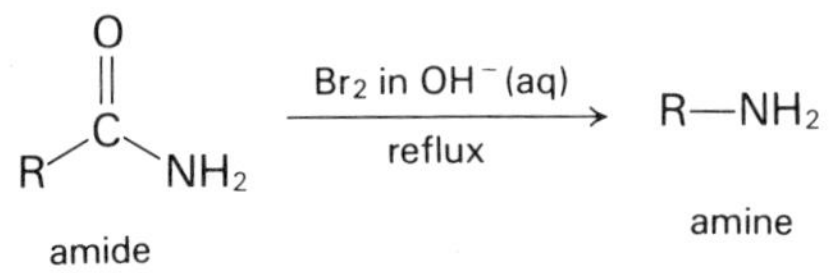

$$\underset{\text{amide}}{R-\underset{\underset{NH_2}{|}}{\overset{\overset{O}{\|}}{C}}} \xrightarrow[\text{reflux}]{Br_2 \text{ in } OH^- \text{(aq)}} \underset{\text{amine}}{R-NH_2}$$

13 Organic Reagents: A Summary

13.1 DIAGNOSTIC REAGENTS TO TEST FOR DIFFERENT GROUPS

Acids

H_3O^+(aq) protonates bases in solution: acid dissolves an amine which is insoluble in water, and protonates the carboxylate anions to produce an insoluble carboxylic acid.

$$R-NH_2 \xrightarrow{H_3O^+(aq)} R-NH_3^+(aq)$$

$$R-COO^-(aq) \xrightarrow{H_3O^+(aq)} R-COOH(s)$$

H_2SO_4(l) protonates even weak bases: for example, concentrated sulphuric acid catalyses ester formation when added to an alcohol and carboxylic acid, and also protonates and then dehydrates methanoates and ethanedioates.

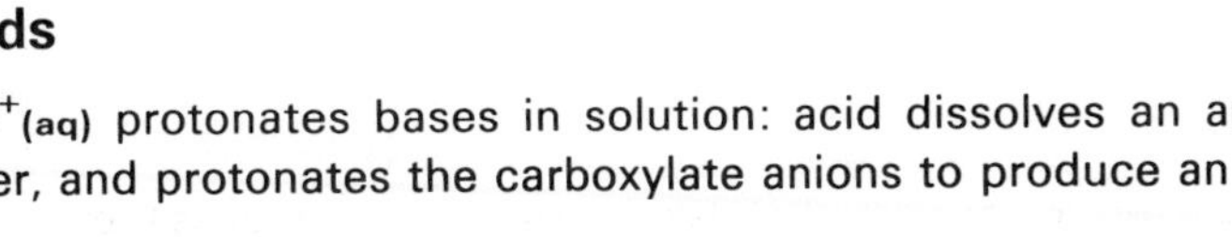

$$H-\underset{O^-}{\overset{O}{\underset{\|}{C}}}-H \xrightarrow{H_2SO_4} \left[H-\underset{O-H}{\overset{O}{\underset{\|}{C}}}-H \right] \xrightarrow{-H_2O} CO$$

$$\underset{^-O}{\overset{O}{\underset{\|}{C}}}-\underset{O^-}{\overset{O}{\underset{\|}{C}}} \xrightarrow{H_2SO_4} \left[\underset{HO}{\overset{O}{\underset{\|}{C}}}-\underset{OH}{\overset{O}{\underset{\|}{C}}} \right] \xrightarrow{-H_2O} CO + CO_2$$

Apart from its acidic and dehydrating properties, sulphuric acid also acts as an oxidant.

HCN gives hydroxynitriles with aldehydes and ketones.

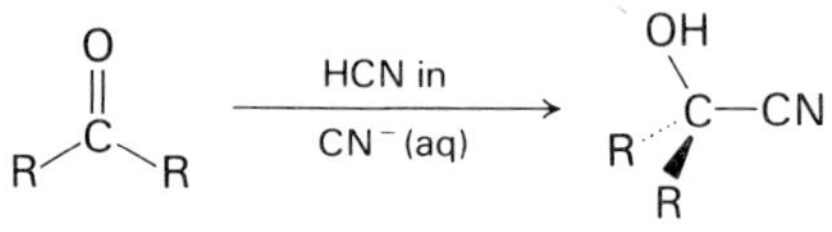

$$R-\underset{R}{\underset{\|}{C}}=O \xrightarrow[CN^-(aq)]{HCN \text{ in}} R-\underset{R}{\overset{OH}{\underset{|}{C}}}-CN$$

HONO (HNO$_2$) converts amines to alcohols with the evolution of nitrogen. Aromatic amines react below 8°C to give diazonium salts.

$$R-NH_2 \xrightarrow{1.\ HCl\ \ 2.\ NaNO_2} R-OH + N_2$$

$$Ar-NH_2 \xrightarrow[T \leq 8°C]{1.\ HCl\ \ 2.\ NaNO_2} [Ar-\overset{+}{N}\equiv N]\ Cl^-$$

Addition of phenols or phenylamines to the diazonium salt produces intensely-coloured dyes.

Bases

OH$^-$(aq) deprotonates acids in solution: for example, alkali dissolves a carboxylic acid or a phenol which is insoluble in water.

$$R-COOH(s) \xrightarrow{OH^-(aq)} R-COO^-(aq)$$

$$Ar-OH(s) \xrightarrow{OH^-(aq)} Ar-O^-(aq)$$

Aqueous alkali is also widely used to cause hydrolysis, for example, of halogenoalkanes, esters, amides and ammonium salts.

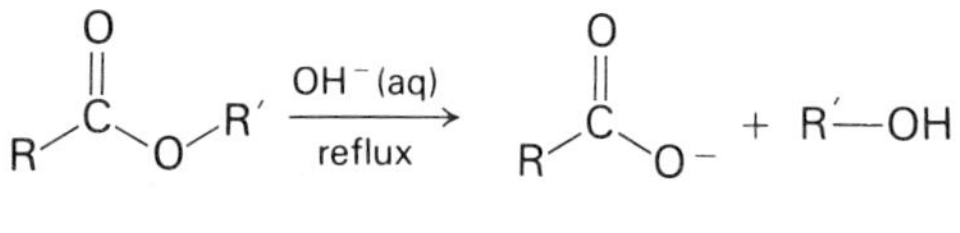

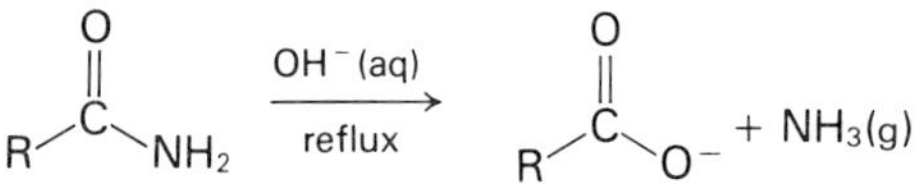

$$R-\underset{\|}{\overset{O}{C}}-O-R' \xrightarrow[\text{reflux}]{OH^-(aq)} R-\underset{\|}{\overset{O}{C}}-O^- + R'-OH$$

$$R-\underset{\|}{\overset{O}{C}}-NH_2 \xrightarrow[\text{reflux}]{OH^-(aq)} R-\underset{\|}{\overset{O}{C}}-O^- + NH_3(g)$$

CO$_3^{2-}$(aq) is used to test for carboxylic acids. An excess of carboxylic acid liberates carbon dioxide from aqueous sodium carbonate solution.

$$R-COOH(s) \xrightarrow{CO_3^{2-}(aq)} R-COO^-(aq) + CO_2(g) + H_2O(l)$$

NH$_3$(aq) is a weak source of hydroxide ions, because of the following equilibrium:

$$NH_3(aq) + H_2O(l) \rightleftharpoons NH_4^+(aq) + OH^-(aq)$$

Concentrated aqueous ammonia also acts as a nucleophilic reagent, and an amide can be produced from an ester.

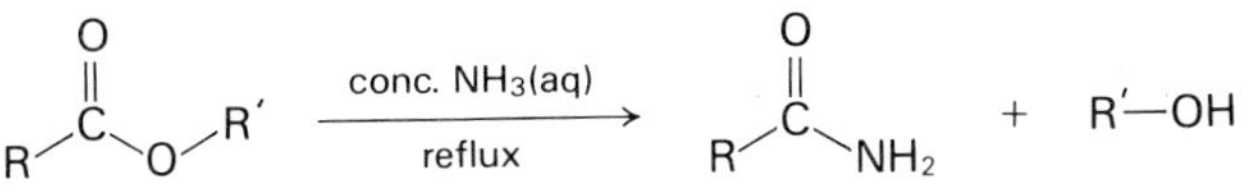

$$\underset{R}{\overset{O}{\|}}\underset{}{C}-O-R' \xrightarrow[\text{reflux}]{\text{conc. NH}_3\text{(aq)}} \underset{R}{\overset{O}{\|}}\underset{}{C}-NH_2 + R'-OH$$

Electrophilic reagents

Br_2(aq) is decolourized by phenols, phenylamines, alkenes and slowly by enols.

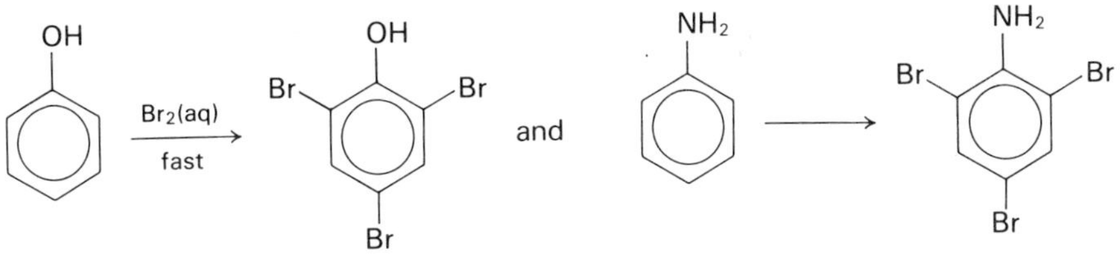

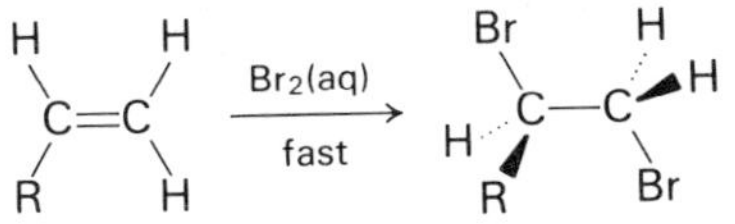

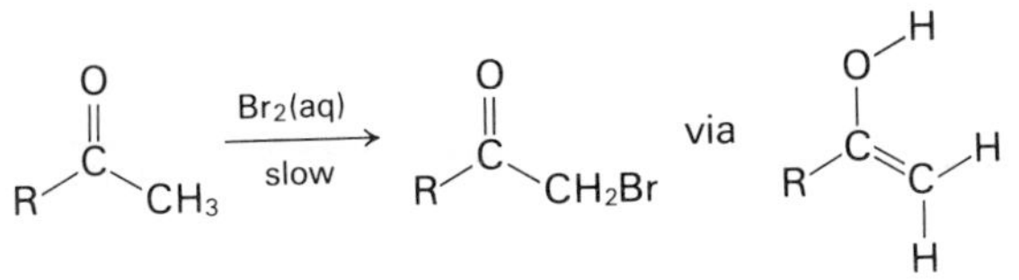

[Phenol + Br₂(aq) fast → 2,4,6-tribromophenol] and [phenylamine → 2,4,6-tribromophenylamine]

[Alkene RCH=CH₂ + Br₂(aq) fast → RCHBr-CH₂Br]

[R-CO-CH₃ + Br₂(aq) slow → R-CO-CH₂Br via enol R-C(OH)=CH₂]

O_3. Ozonolysis is fully described on page 192. It is used to characterize substances containing carbon–carbon double bonds.

I_2 in NaOH. Iodoform is precipitated when iodine in alkali is added to a drop of test material whose molecules contain the groups shown below.

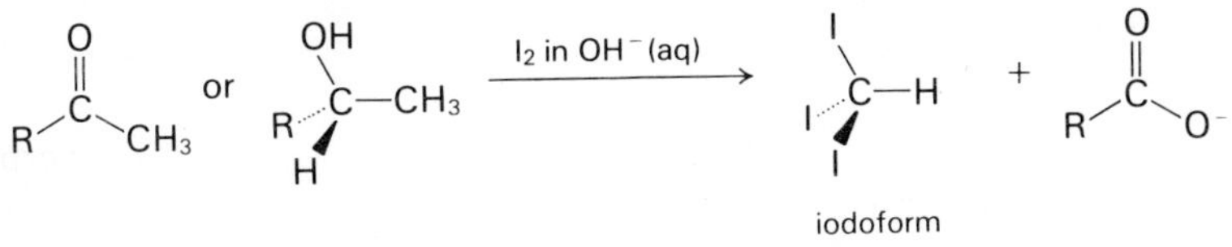

$$\underset{R}{\overset{O}{\|}}\underset{}{C}-CH_3 \text{ or } \underset{R}{\overset{OH}{\underset{H}{|}}}\underset{}{C}-CH_3 \xrightarrow{I_2 \text{ in OH}^-\text{(aq)}} \underset{I}{\overset{I}{\underset{I}{|}}}\underset{}{C}-H + \underset{R}{\overset{O}{\|}}\underset{}{C}-O^-$$

iodoform

Metal ions

Fe^{3+}(aq). Aliphatic acids, aromatic acids and phenols give different characteristic colours with aqueous iron(III) ions. These are shown overleaf.

$$\text{R—COOH} \xrightarrow{Fe^{3+}(aq)} \text{red complex}$$

$$\text{Ar—COOH} \xrightarrow{Fe^{3+}(aq)} \text{buff complex}$$

$$\text{Ar—OH} \xrightarrow{Fe^{3+}(aq)} \text{purple complex}$$

Cu^{2+}(aq). Aliphatic amines, aromatic amines and conjugated polypeptides (see page 229) give different characteristic colours with aqueous copper(II) ions.

$$\text{R—NH}_2 \xrightarrow{Cu^{2+}(aq)} \text{blue complex}$$

$$\text{Ar—NH}_2 \xrightarrow{Cu^{2+}(aq)} \text{green complex}$$

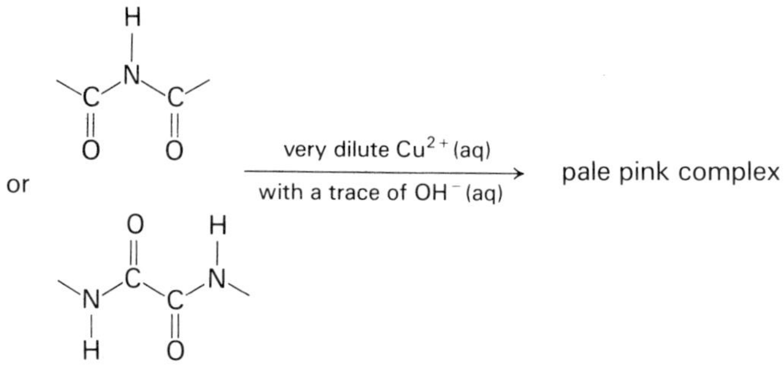
or
very dilute Cu2+(aq)
with a trace of OH−(aq)
pale pink complex

Nucleophilic reagents

Apart from ammonia and sodium hydroxide (which are described earlier), hydroxylamine and 2,4-dinitrophenylhydrazine (2,4-dnph) are nucleophilic reagents used in diagnostic tests. They confirm the presence of carbonyl compounds.

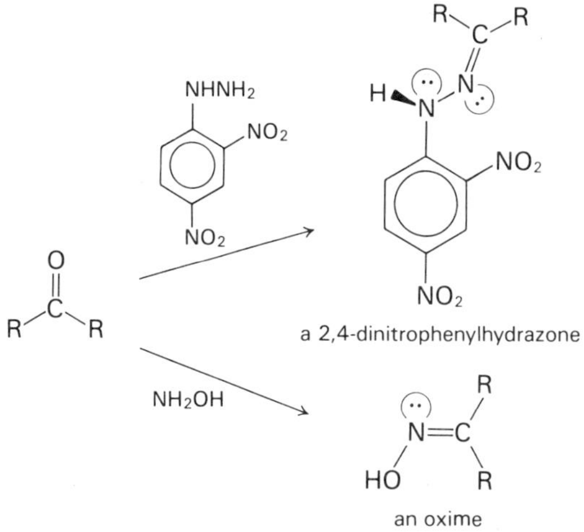
NHNH2
NO2
NO2
NH2OH
a 2,4-dinitrophenylhydrazone
an oxime

Redox reagents

MnO$_4^-$(aq) is decolourized in acid solution by primary and secondary alcohols, aldehydes and alkenes. Under alkaline conditions, the purple manganate(VII) is converted to a brown precipitate of manganese(IV) oxide by aldehydes and alkenes.

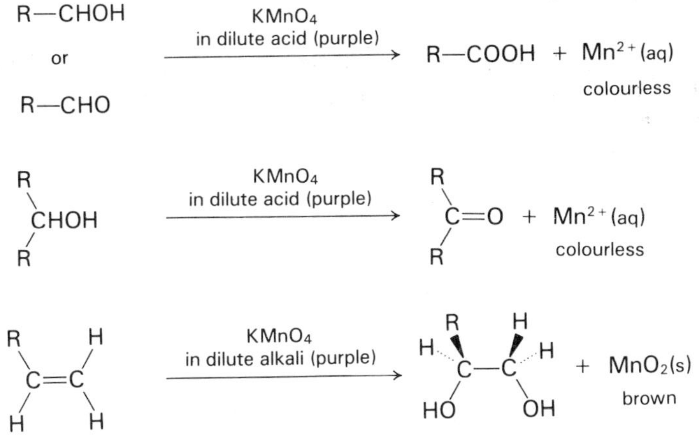

$$\text{R—CHOH or R—CHO} \xrightarrow[\text{in dilute acid (purple)}]{\text{KMnO}_4} \text{R—COOH} + \text{Mn}^{2+}\text{(aq) colourless}$$

$$\text{R}_2\text{CHOH} \xrightarrow[\text{in dilute acid (purple)}]{\text{KMnO}_4} \text{R}_2\text{C=O} + \text{Mn}^{2+}\text{(aq) colourless}$$

$$\text{H}_2\text{C=CH}_2 \xrightarrow[\text{in dilute alkali (purple)}]{\text{KMnO}_4} \text{HOCH}_2\text{—CH}_2\text{OH} + \text{MnO}_2\text{(s) brown}$$

Alkaline manganate(VII) is usually used to test for carbon–carbon double bonds.

Ammoniacal silver nitrate. When ammonia is added to silver nitrate, a precipitate of silver oxide forms but redissolves in excess ammonia. Aldehydes are able to reduce the complexed silver(I) ions to a mirror of silver on the inside of the tube.

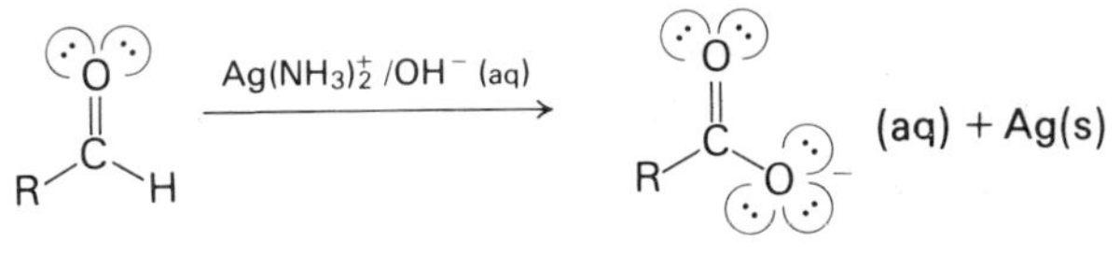

$$\text{RCHO} \xrightarrow{\text{Ag(NH}_3)_2^+ / \text{OH}^- \text{(aq)}} \text{RCO}_2^- \text{(aq)} + \text{Ag(s)}$$

13.2 SYNTHETIC REAGENTS

Reagents for aliphatic functional groups

Acid chlorides and anhydrides are used to make esters and amides. Ethanoylation and benzoylation are particular examples of their action as synthetic reagents. For example, phenylamine is benzoylated to N-phenyl benzamide as shown below.

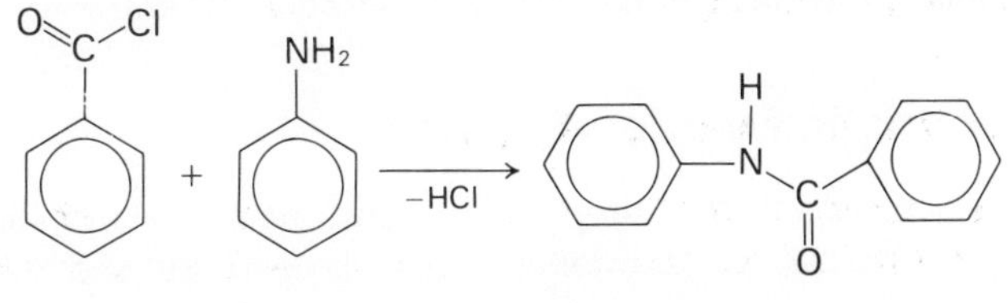

$$\text{C}_6\text{H}_5\text{COCl} + \text{C}_6\text{H}_5\text{NH}_2 \xrightarrow{-\text{HCl}} \text{C}_6\text{H}_5\text{-NH-CO-C}_6\text{H}_5$$

PCl$_5$ and SOCl$_2$ convert a hydroxyl group to a chlorogroup. Sulphur dichloride oxide is often the more useful synthetic reagent because the by-products are all gaseous.

$$R-OH(l) \xrightarrow{PCl_5} R-Cl(l) + POCl_3(l) + HCl(g)$$

$$R-OH(l) \xrightarrow{SOCl_2} R-Cl(l) + SO_2(g) + HCl(g)$$

KCN provides a means by which another carbon atom can be added on to the structure of a compound. Cyanide ions displace bromide ions from a bromoalkane, and also add to ketones and aldehydes.

$$R-Br \xrightarrow[\text{reflux}]{CN^-(aq)} R-CN$$

$$\underset{R}{\overset{O}{\underset{\|}{R-C-R}}} \xrightarrow{HCN \text{ in } CN^-(aq)} R-\underset{R}{\overset{OH}{\underset{|}{C}}}-CN$$

Since nitriles are readily hydrolysed to acids under either alkaline or acidic conditions (see page 227), the above process can be a useful half-way stage during the synthesis of a substituted acid.

HONO (HNO$_2$) converts an amino group into a hydroxyl group.

$$R-NH_2 \xrightarrow{HONO} R-OH + N_2 + H_2O$$

Redox reagents. Lithium tetrahydridoaluminate hydrogenates polar unsaturated groups but does not reduce alkenes.

$$\overset{\delta-O}{\underset{R}{\overset{\|}{\underset{\delta+}{C}}}-R} \xrightarrow[\text{2. } H_2O]{\text{1. LiAlH}_4} R-\overset{OH}{\underset{R}{\overset{|}{C}}}-H \quad \text{and} \quad \overset{\delta-N}{\underset{R}{\overset{\|\|}{\underset{\delta+}{C}}}} \xrightarrow[\text{2. } H_2O]{\text{1. LiAlH}_4} H-\overset{NH_2}{\underset{H}{\overset{|}{C}}}-R$$

Hydrogen and a nickel or palladium catalyst hydrogenate the non-polar saturated groups present in alkenes and alkynes. Carbonyls are much more resistant to hydrogenation, but nitriles can be reduced at high pressure in the presence of ammonia.

Potassium dichromate(VI) and manganate(VII) are used as oxidants to oxidize primary alcohols to acids and aldehydes, and secondary alcohols to ketones.

Reagents for aromatic functional groups

HNO$_3$/H$_2$SO$_4$ is a source of the nitryl (nitronium) cation which is an electrophile. Aromatic rings are nitrated at positions which depend on the nature of the substituents already present.

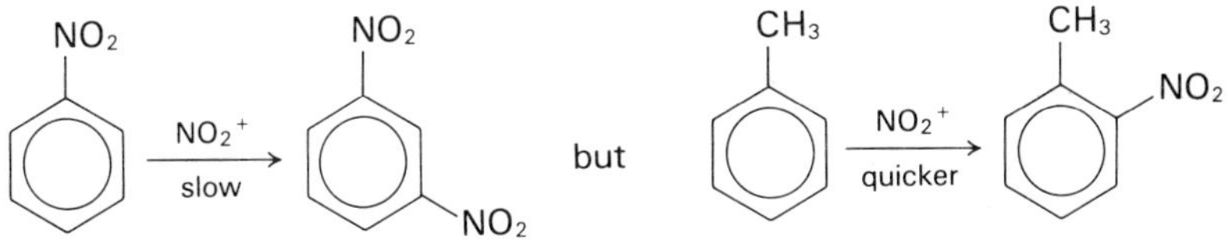

[Diagram: nitrobenzene + NO₂⁺ → 1,3-dinitrobenzene (slow), but methylbenzene + NO₂⁺ → 2-nitromethylbenzene (quicker)]

R—Cl or R—COCl (in the presence of aluminium chloride); these are known as Friedel–Crafts reagents. They provide a source of carbonium ions as electrophiles. For example,

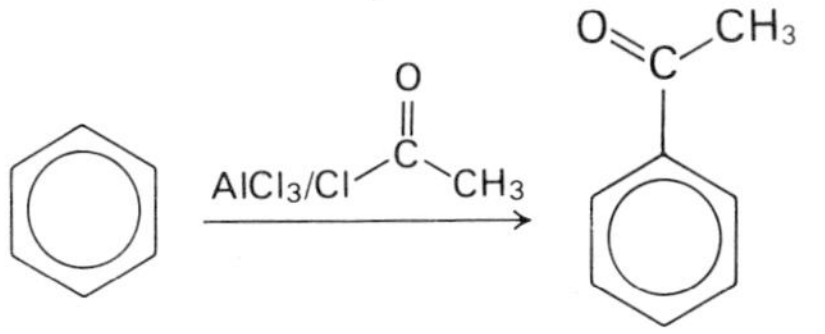

[Diagram: benzene + CH₃COCl/AlCl₃ → phenyl methyl ketone (acetophenone)]

A diazonium salt is a useful half-way stage in the synthesis of different aromatic compounds. A full discussion can be found on page 208.

Redox reagents. An aromatic nitro-group is reduced to an amine by tin and concentrated hydrochloric acid. This often provides the first step in the synthesis of a diazonium salt.

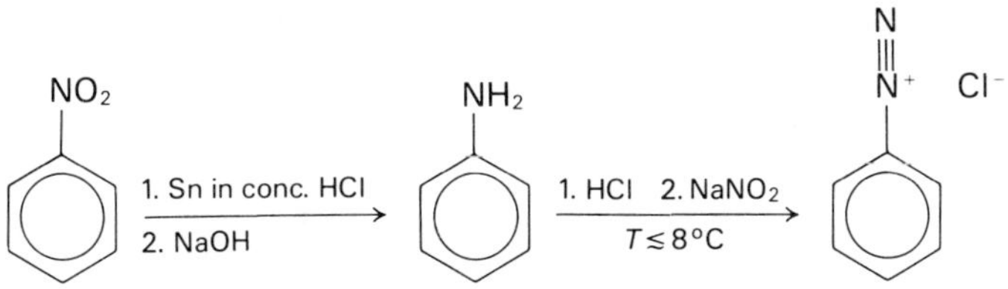

[Diagram: nitrobenzene →(1. Sn in conc. HCl, 2. NaOH)→ aniline →(1. HCl, 2. NaNO₂, T ≤ 8°C)→ benzenediazonium chloride]

Oxidation of an aromatic side-chain can be achieved without breaking up the aromatic ring. Refluxing in acidified dichromate(VI) or manganate(VII) converts any side-chain to a carboxylic acid group. For example,

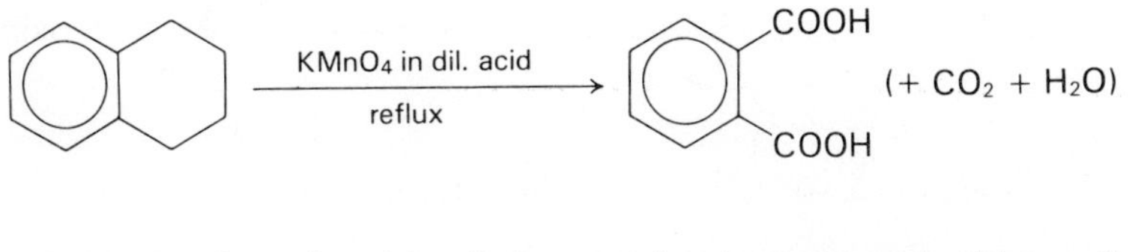

[Diagram: tetralin →(KMnO₄ in dil. acid, reflux)→ benzene-1,2-dicarboxylic acid (+ CO₂ + H₂O)]

Free radical chlorination of a side-chain can also be achieved without disturbing the aromatic ring. As long as the process is carried out in the liquid phase, good yields are produced. Methylbenzene can be oxidized in this way to benzyl alcohol (phenylmethanol), benzaldehyde and benzoic acid.

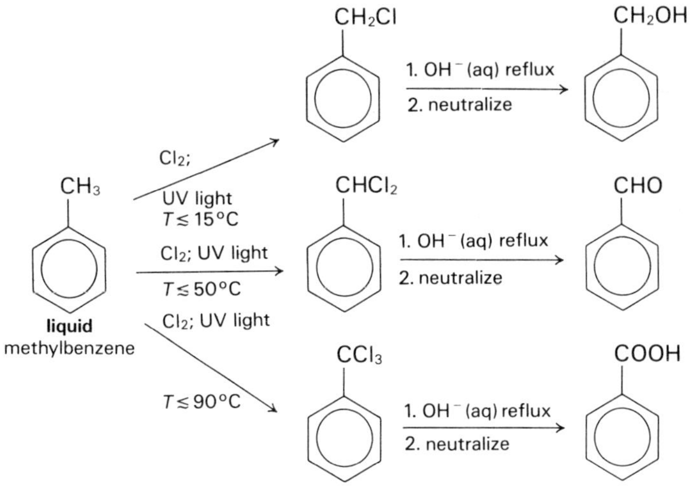

[Reaction scheme: liquid methylbenzene (C$_6$H$_5$CH$_3$) undergoes side-chain chlorination with Cl$_2$ and UV light at different temperatures to give three chlorinated products, which are then hydrolysed:

- Cl$_2$; UV light, $T \leq 15°C$ → C$_6$H$_5$CH$_2$Cl → (1. OH$^-$(aq) reflux; 2. neutralize) → C$_6$H$_5$CH$_2$OH
- Cl$_2$; UV light, $T \leq 50°C$ → C$_6$H$_5$CHCl$_2$ → (1. OH$^-$(aq) reflux; 2. neutralize) → C$_6$H$_5$CHO
- Cl$_2$; UV light, $T \leq 90°C$ → C$_6$H$_5$CCl$_3$ → (1. OH$^-$(aq) reflux; 2. neutralize) → C$_6$H$_5$COOH]

14 Stereochemistry

14.1 ISOMERISM

ISOMERS
are different compounds which have the same molecular formula

- **STRUCTURAL ISOMERS**
  have the same molecular formula but different structural formulas
- **STEREOISOMERS**
  have the same structural formula but different arrangements of atoms in space
  - **GEOMETRIC ISOMERS**
    the difference in arrangement is due to a restriction on the rotation of bonds within the structure
  - **OPTICAL ISOMERS**
    the difference in arrangement is due to a lack of planes or centres of symmetry within the structure

Structural isomers

Two different molecules often have the same number of the same atoms bonded together. For example, methoxymethane and ethanol are both C_2H_6O, but the order in which the atoms are linked is different in the two molecules.

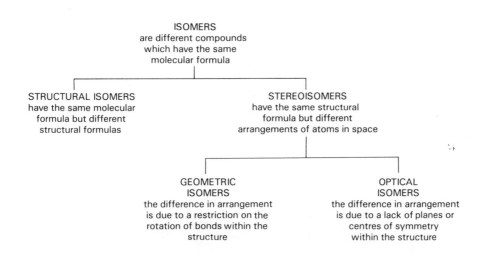

methoxymethane   ethanol

And there are three different forms of hydrated chromium(III) chloride, $CrCl_3 \cdot 6H_2O$.

$$\left[\begin{array}{c} H_2O \\ \downarrow \\ H_2O \rightarrow Cr \leftarrow OH_2 \\ H_2O \nearrow \uparrow \\ H_2O \end{array}\right]^{3+} (Cl^-)_3 \quad \left[\begin{array}{c} H_2O \\ \downarrow \quad Cl \\ H_2O \rightarrow Cr \leftarrow OH_2 \\ H_2O \nearrow \uparrow \\ H_2O \end{array}\right]^{2+} (Cl^-)_2 \quad \left[\begin{array}{c} H_2O \\ \downarrow \quad Cl \\ H_2O \rightarrow Cr \leftarrow Cl \\ H_2O \nearrow \uparrow \\ H_2O \end{array}\right]^{+} Cl^-$$

violet   pale green   dark green

These are examples of structural isomerism.

There are two types of structural isomerism which are of particular importance: positional isomerism and tautomerism.

Positional isomers are the structural isomers of cyclic compounds. For example, there are three structural isomers of dinitrobenzene.

1,2-dinitrobenzene   1,3-dinitrobenzene   1,4-dinitrobenzene

Tautomers are structural isomers in dynamic equilibrium with each other. The most important example of tautomerism is *keto–enol* tautomerism (see page 211). This occurs in carbonyl compounds containing α-hydrogen atoms. For example, propanone exhibits tautomerism.

keto-form   *enol*-form   *keto*-form

Stereoisomers

Although two molecules may contain the same atoms bonded in the same order, it is still possible that the arrangement of atoms may not be the same.

Geometric isomers differ in the position (with respect to each other) of a pair of substituents bonded to different parts of the molecule. In one isomer, the pair are held on the same side of the molecule: in the other isomer, the pair are on opposite sides.

This is caused by a restriction on the internal rotation of the part of the molecule linking the two substituents. For example π-overlap prevents the rotation of the double-bonded carbon atoms below.

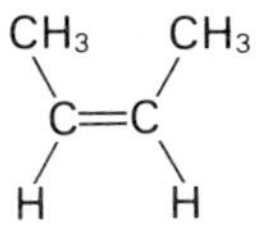

cis-but-2-ene 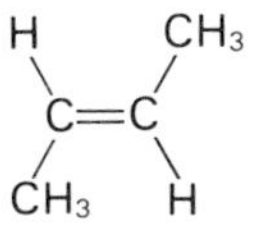   trans-but-2-ene

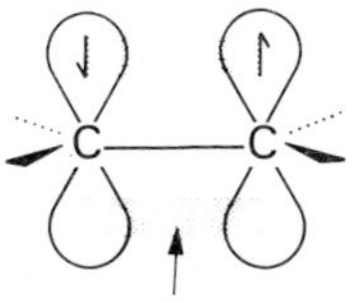

π-overlap of the 2p-orbitals restricts the rotation about the bonded carbon atoms

In the first molecule, the methyl groups are on the same side: *cis*-but-2-ene. In the second molecule, the two groups are on opposite sides: *trans*-but-2-ene.

Saturated ring structures and square planar or octahedral complexes also exhibit geometric (or *cis–trans*) isomerism. Some examples are shown below.

1,2-dibromo-cyclopropane	diammineplatinum(II) chloride	potassium bis(ethanediato)-chromate(III)-2-water
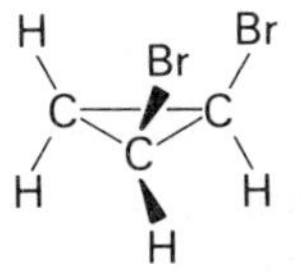 cis-	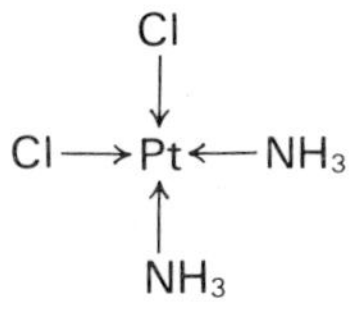 cis-	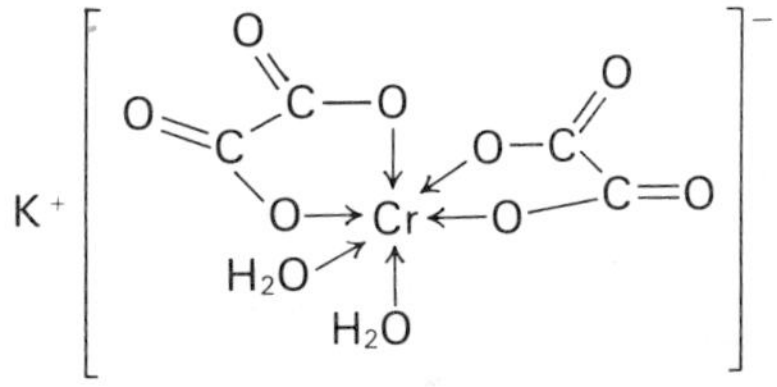 cis-
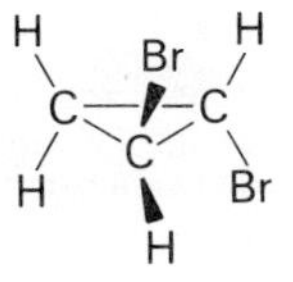 trans-	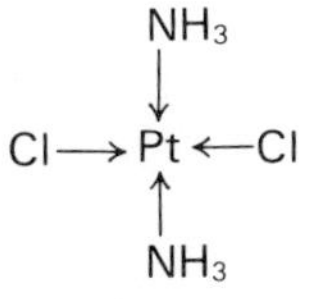 trans-	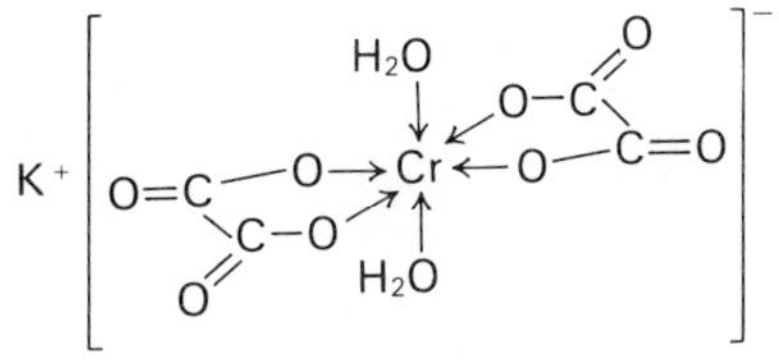 trans-

Optical isomers do not depend on restricted rotation for their distinction. If a structure has no centre nor plane of symmetry, it is possible to have two forms of the structure that are non-superimposable mirror images of each other. The two forms have an identical structural formula and are free to rotate about each bond, but cannot be superimposed on each other no matter how the atoms are rotated. These two different stereoisomers are called enantiomers. For example,

242 Condensed Chemistry

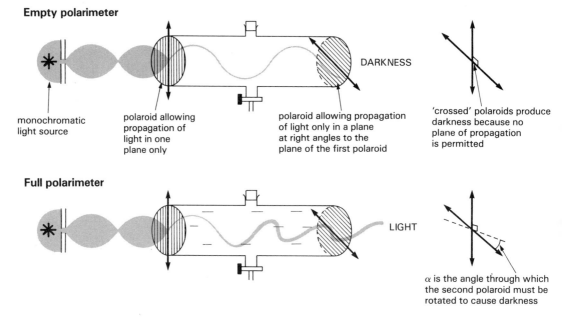

mirror

If the mirror image is lifted out of the paper and turned through 180°, its methyl group and carboxylic acid group superimpose on those of the first structure. However, the other two groups do not match up. If the other two are now made to match up, it is found that the methyls and acid groups are no longer superimposed.

The effect is known as 'optical' isomerism because a solution containing a pure enantiomer interacts with a beam of polarized light passing through it. The plane of the polarization is rotated through an angle of up to about 30°. The angle is measured by a polarimeter, which has a small transparent cell separating two 'crossed' polaroids, one of which can be turned with respect to the instrument.

When full of a solution of an enantiomer, the plane of the propagation of the polarized light is rotated slightly. This means that the position of the second polaroid is no longer 'crossed' with respect to the polarized light reaching it. In order to prevent the passage of light through the polaroid, it must be turned through an angle equal to the angle of rotation of the plane of the polarized light. It is found that the angles measured for a pair of enantiomers are exactly equal, but in opposite directions. Hence, one enantiomer is called the (+) form and the other the (−) form; an equimolar mixture of the two is known as a racemic mixture (±).

Most enantiomers are based on tetravalent carbon atoms (like those shown at the top of this page). When there are four different groups bonded to a carbon atom, the structure has no centre nor plane of symmetry. The carbon atom is an 'asymmetric centre'. Substituted phosphines and certain transition metal complexes are among the examples of inorganic enantiomers.

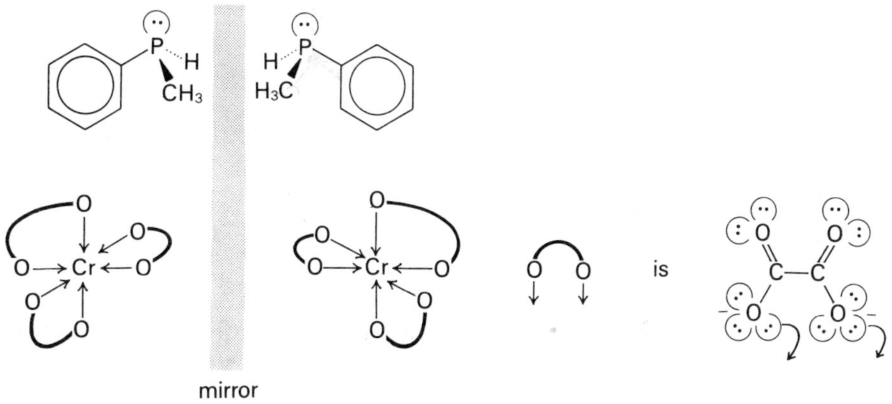

Many organic molecules contain more than one asymmetric carbon atom. Paradoxically, the bonding together of two centres of equivalent asymmetry produces a plane of symmetry and, therefore, a lack of optical properties. For example, there are only three stereoisomers of 2,3-dibromobutane:

(+) 2,3-dibromobutane 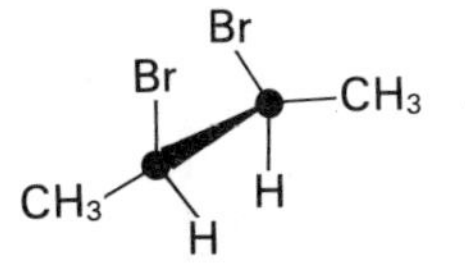	(−) 2,3-dibromobutane 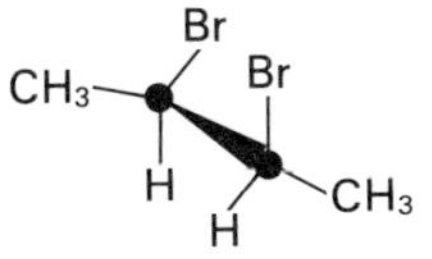	meso-2,3-dibromobutane 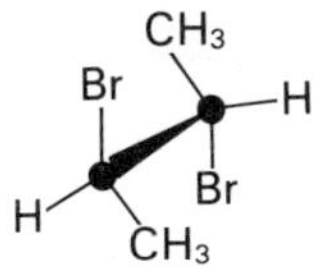
or, looking along the line of the carbon-carbon bond,	or, looking along the line of the carbon-carbon bond	or, looking along the line of the carbon-carbon bond

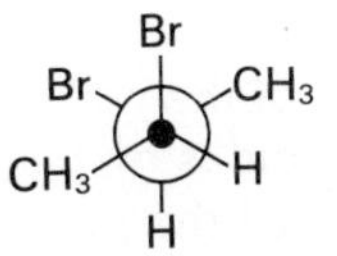

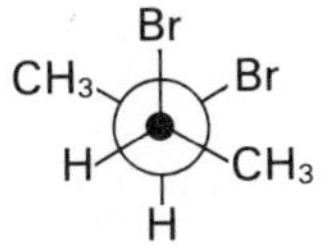

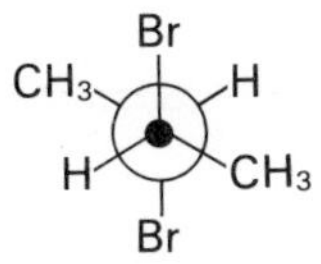

The first two are non-superimposable mirror images of each other and are, therefore, enantiomers. The third also cannot be superimposed on either of the first two, but it is a mirror image of *neither*. It is a 'diastereoisomer' of the first two, and it does not show any optical properties. This suggests that it has a centre or plane of symmetry, despite possessing two asymmetric centres. When the front carbon atom is rotated through

180°, it becomes evident that the two halves of the molecule are mirror images of each other.

The molecule is 'internally' compensated in the sense that racemic mixtures are 'externally' compensated. This third stereoisomer is called the *meso-* form. It is, therefore, possible to have (+), (−) and *meso-* 2,3-dibromobutane as well as (±) 2,3-dibromobutane.

14.2 STEREOSPECIFICITY AND STEREOSELECTIVITY

Factors that determine the course of a reaction

In many organic reactions, there is more than one possible set of products. Frequently, however, despite the availability of an alternative pathway, one particular set of products dominates. A reaction in which those products are generated to the exclusion of all others is said to be stereospecific. If other products *are* formed, but in relatively tiny amounts, the process is stereoselective. In general, the causes of stereospecificity or stereoselectivity depend upon the relative rates of the competing processes. The reaction follows the pathway that happens most quickly. In other words, the reaction proceeds via the activated complex of lowest energy. Compare the energy profiles of the two competing processes on the following energy diagram.

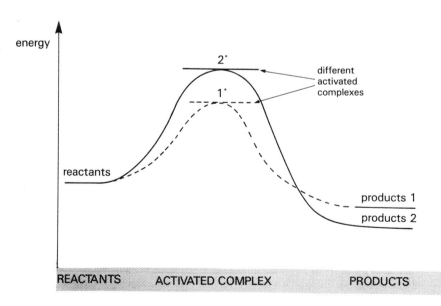

Products (2) are more stable than products (1) and it might, therefore, be expected that they would be produced in preference to products (1). However, the relative stability of the activated complexes, 1* and 2*, shows that the activation energy of the first process is considerably less than that of the second. The rate of the first process is, therefore, much greater, and so the products are controlled by 1*. The reaction is said to be kinetically controlled. Kinetic control can only be upset by allowing the reaction mixture to reach a position of equilibrium over a long period of time. Under these conditions, a reaction is said to be thermodynamically controlled. However, the vast majority of organic reactions are kinetically controlled.

The formation of an activated complex. There are two main factors that control the formation of a particular activated complex:

1 its energy compared with that of any other possible complexes;
2 the steric requirements for its formation.

The relative energy of a complex is assessed by inspecting the ability of the structure to delocalize charge. Overlap drawings, or a consideration of canonical forms, can be used to achieve this. For example, when an electrophile (E^+) adds to phenol there are three possible σ-complexes depending on the position of the addition. It can be shown that the 2-substituted complex is of lower energy than the 3-substituted complex because there are a greater number of canonical forms of the 2-substituted complex.

four canonical forms of the 2-substituted complex

three canonical forms of the 3-substituted complex

Occasionally, the formation of the most stable complex is sterically hindered. For example, although chloroethane is hydrolysed by refluxing in alkali (below), steric hindrance is largely responsible for the inertness of tetrachloromethane in alkali.

In the case of tetrachloromethane, the formation of a similar activated complex is hindered by the presence of the three bulky chlorine atoms. The attack of the lone pair on the carbon atom is considerably restricted.

Examples

Nucleophilic substitution. This may occur by two possible mechanisms, S_N1 or S_N2 (see page 182). S_N2 dominates unless the nucleophile is bulky, or the carbon atom to be attacked has bulky groups attached to it. S_N1 can contribute significantly either if S_N2 is sterically hindered (as described above), or if the carbonium ion produced is comparatively stable. For example, in the hydrolysis of phenylbromomethane:

the phenyl group hinders the approach of an attacking nucleophile to the carbon atom

four canonical forms of the S_N1 carbonium ion

The S_N2 mechanism leads to an inversion of the configuration of a saturated carbon atom because the attacking nucleophile pushes the three substituents into the same plane as the carbon atom during the formation of the activated complex. On the departure of the leaving group, these three groups move further away from the nucleophile. The effect is like an umbrella blowing out in the wind.

Optical activity is, therefore, preserved if an enantiomer undergoes nucleophilic substitution via an S_N2 mechanism. This is *not* so for an S_N1 mechanism because the nucleophile can attack a planar carbonium ion from either side. So, for example (+) 1-phenyl-1-bromoethane is hydrolysed to (±) 1-phenylethanol; the bromoderivative is racemized, see page 242.

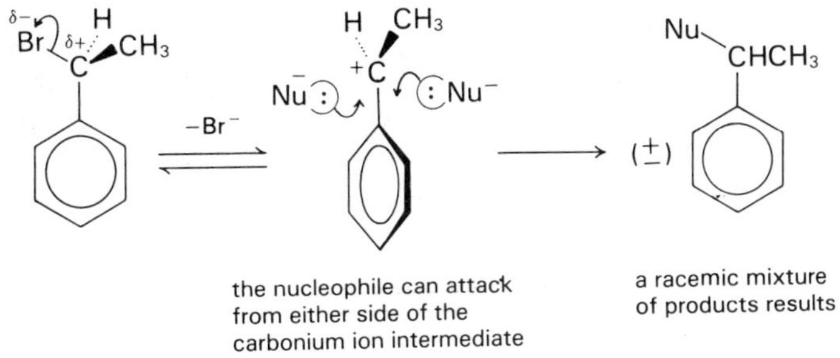

the nucleophile can attack from either side of the carbonium ion intermediate

a racemic mixture of products results

When the conditions for nucleophilic substitution are unfavourable for any of the reasons outlined above, elimination may become an important competing process. For example, ethanol is produced when bromoethane is refluxed with dilute sodium hydroxide, but ethene is the major product when the sodium salt of methylpropan-2-ol is used.

substitution 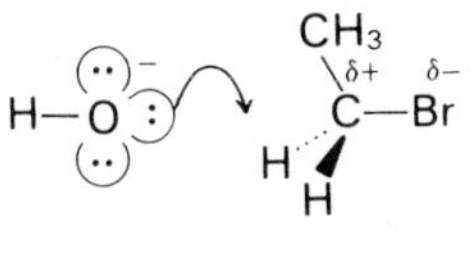	elimination 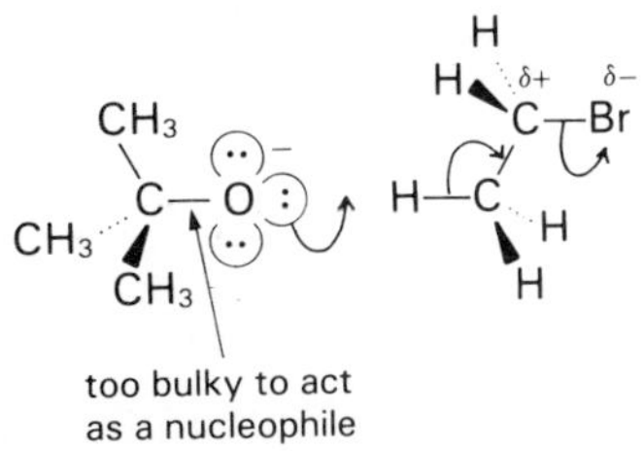

too bulky to act as a nucleophile

Electrophilic substitution. A substituent on an aromatic ring directs electrophilic attack to particular positions on the ring. The formation of stereoselective products can be explained in terms of the relative stability of the possible activated complexes. The case of phenol is discussed on page 245: substitution at 2-, 4- and 6- is more favourable than at 3- and 5- because there are a greater number of canonical forms of a 2-, 4- or 6- substituted complex.

The argument can be generalized further by considering the possible sites where the positive charge is localized on an aromatic σ-complex.

position of electrophile (E) with respect to substituent (X)	possible canonical forms
2- or 6-substituted complexes	
4-substituted complex	
3- or 5-substituted complexes	

The positive sites can be shown on a single hexagon each. For the 2-, 4- and 6- substituted positions, the sites occupy the corners of a triangle connected to the substituent X. For the 3- and 5- substituted positions, they occupy the corners of a different triangle.

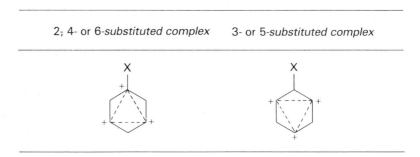

A substituent X that exerts a +*M* or +*I* effect (see page 181) stabilizes the 2-, 4- and 6- substituted complexes because these each have one canonical form in which X bonds directly to the carbon atom with a localized positive charge. The 3- and 5- substituted complexes do *not* have a canonical form of this sort.

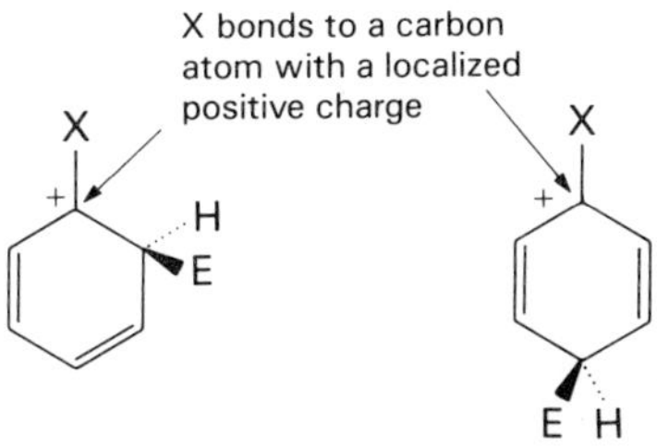

However, if the substituent exerts a $-M$ or $-I$ effect, it destabilizes the canonical form shown above. Under these conditions, the 3- and 5- substituted complexes are *more* stable, and substitution at the 2-, 4- and 6- positions is less likely.

Electrophilic addition: Markovnikov's rule. When a strong acid adds to an asymmetric alkene, it is found that the hydrogen atom bonds to the carbon atom with the most hydrogen atoms already attached. This is known as Markovnikov's rule (see page 191). The relative stabilities of the possible σ-complexes are responsible for this effect. For example, when hydrogen bromide adds to propene, the π-complex produced can break down into two possible σ-complexes:

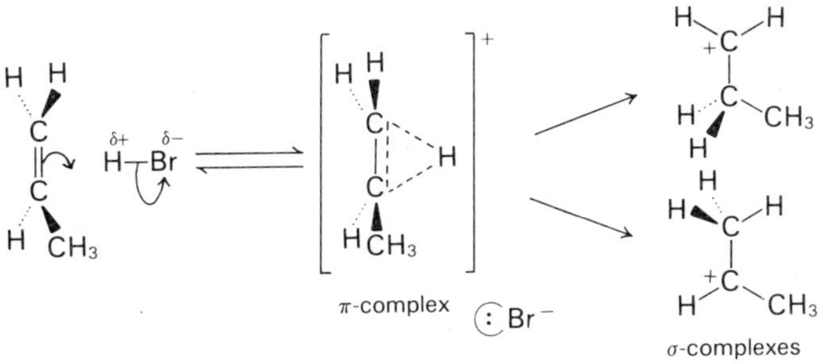

In the first σ-complex, the carbonium ion contains two hydrogen atoms and one alkyl group; in the second, there are two alkyl groups and one hydrogen atom. Since an alkyl group exerts a $+I$ effect compared with a hydrogen atom, the positive charge is stabilized better by the second σ-complex, and hence the products are derived from this complex.

The positive inductive effect of an alkyl group is due to combined 'secondary' effects. For example, a methyl group has three hydrogen atoms and the carbon atom transmits charge drawn from all three:

(ii)

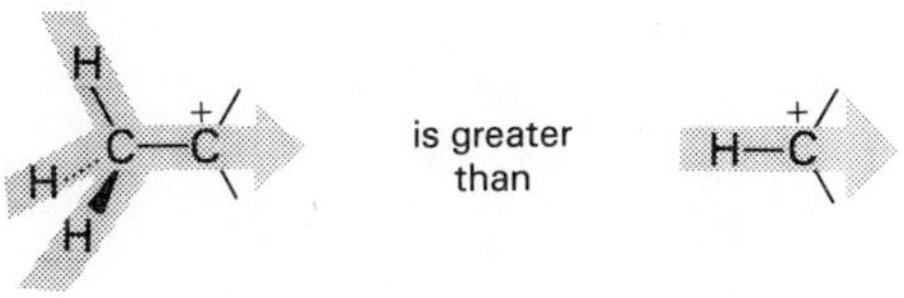

Index

A-metal 126
α-aminoacid 225
α-hydrogen atom 211, 214
α-particle 1, 3
acid 84, 170, 218
  catalysis 77, 215, 227
  conjugate 85
  dissociation constant K_a 86
  strength 86, 91, 161, 219
activation energy E_A 53, 73, 108
  catalysed reaction 76
activated complex 76, 179, 185, 244–9
addition reaction
  electrophilic 185, 191, 249
  nucleophilic 183, 212
adsorption 78
alcohols 200–5
aldehydes 210–7
aldol reaction 217
alkali 110, 162, 168
alkanes 189–90
alkenes 190–4
allotropy 51
  dynamic 53
  enantiotrophy 51, 136, 152
  monotropy 52, 137, 152
aluminates 115, 166
aluminium 23–4, 114–7, 166
amides 224–30
amines 205–10
ammonia decomposition 78
amphoteric behaviour 85, 116, 141
Andrew's curves 33
anti-bonding orbital 15, 145
aromaticity 66, 194–7
Arrhenius equation 74
asymmetric centre 242
atomic
  mass A_r 20, 22
  number A 2, 109
  orbital 8
  spectrum 4
aufbau principle 9
Avogadro's
  constant L 20, 29
  hypothesis 29, 32, 190
axial position 17
azeotrope 37

B-metal 126
β-particle 3
base 84, 140
  catalysis 217, 227
  conjugate 85
  dissociation constant K_b 86
  strength 86, 91, 148
benzoylation 223
biuret test 229
body-centred cubic packing 27
boiling point T_b 34
  elevation 40

bond 12
  angle 149
  covalent 14, 65
  dissociation energy D^θ 13, 14, 65, 130
  hydrogen 20, 51, 133
  ionic 23, 27, 64
  metallic 23, 25
  pi (π-) 15
  rotation 241
  sigma (σ-) 15
bonding orbital 15
boranes 160
Born–Haber cycle 64
Boyle's law 32
Bragg equation 46
bromine 130–5, 191
Brønsted–Lowry theory 84
brown ring test 151
buffer solutions 89

caesium chloride structure 28
calcium 112–4
calorimeter 59, 60
canonical form 67, 162, 195, 245–8
carbanion 217
carbides 153–4
carbon 152–8, 166
carbonium ion 182, 246, 249
carboxylic acid 218–21
  anhydride 221–3
  chloride 221–3
catalyst 76, 77, 83
catenation 138, 155
cation size 16, 23, 87, 116
chain reaction 78, 189
Charles' law 32
charge separation 19, 181
chemical equilibrium 80
chlorates 132, 134
chlorides 23–4, 131–5
chlorine 79, 130–5, 165, 169, 238
chromates 124
chromium 122–5
clays 156
close packing 25
coal 171
colligative properties 40
collision theory 73
complexes
  colour 121
  formation 120, 128
  isomerism in 241, 243
concentration 69, 80
  effect on conductivity 103, 104
  effect on equilibrium 82
  effect on rate 69, 70
condensation reaction 217, 225
conductivity 100, 102–104
conjugate
  acid-base pair 85
  redox pair 97
conjugation 191
constant boiling mixture 40
contact process 170
continuum 4, 7
cooling
  correction 59
  curve 47
co-ordination number 28

copper 59, 127
core charge 110
coupling reaction 210
covalency 12, 14
  degree of 16, 24, 114
cracking 172
critical temperature T_c 50
cryoscopic constant 41
crystal field 121
cubic close packing 26
cumene-phenol process 175

d-orbital 9, 118
d-splitting 121
Dalton's law 31
datum line 62
decay constant λ 73
degree of
  covalency 16, 24, 114
  dissociation α 101
delocalization energy 66
desorption 78
deviation
  from gas ideality 32
  from Raoult's law 37
diagnostic reagents 231
diamond structure 152
diaphragm cell 169
diastereoisomer 243
diazotization 208–9
dichromates 124–5
diffraction
  grating 4
  X-rays 45
dipole-dipole force 19
discharge
  potential 107–8
  tube 4
disorder 167
disproportionation 97
  chlorine 132
  copper 127
  nitrogen 149
  phosphorus 147
dissociation constant
  electrolytic k 100
  K_a K_b K_w 86
distillation 38
distribution
  constant k 57
  of collision energies 74
  of molecular speeds 31
double bond 15, 162
double salt 148
Downs cell 165
dye 210
dynamic allotropy 53

ebullioscopic constant 40
electrode
  inert 94
  potential E 93
  standard potential $E^\ominus$ 93, 99
electrolysis 106
electrolyte 100
electron 2
  affinity 12
  delocalized 23, 152
  transfer 13, 92